ROBOTIC INDUSTRIALIZATION

Automation and Robotic Technologies for Customized Component, Module, and Building Prefabrication

The Cambridge Handbooks on Construction Robotics series focuses on the implementation of automation and robot technology to renew the construction industry and to arrest its declining productivity. The series is intended to give professionals, researchers, lecturers, and students basic conceptual and technical skills and implementation strategies to manage, research, or teach the implementation of advanced automation and robot technology–based processes and technologies in construction. Currently, the implementation of modern developments in product structures (modularity and design for manufacturing), organizational strategies (just in time, just in sequence, and pulling production), and informational aspects (computer-aided design/manufacturing and computer-integrated manufacturing) are lagging because of the lack of modern integrated machine technology in construction. The Cambridge Handbooks on Construction Robotics books discuss progress in robot systems theory and demonstrate their integration using real systematic applications and projections for off-site as well as on-site building production.

In this volume, concepts, technologies, and developments in the field of building-component manufacturing – based on concrete, brick, wood, and steel as building materials and on large-scale prefabrication, which holds the potential to deliver complex, customized components and products – are introduced and discussed. Robotic industrialization refers to the transformation of parts and low-level components into higher-level components, modules, and finally building systems by highly mechanized, automated, or robot-supported industrial settings in structured off-site environments. Components and modules are (in modular building product structures) independent building blocks that are delivered by suppliers to original equipment manufacturers such as large-scale prefabrication companies or automated/robotic on-site factories. In particular, Japanese large-scale prefabrication companies have altered the building structures, manufacturing processes, and organizational structures significantly to be able to assemble in their factories high-level components and modules from Tier-1 suppliers into customized buildings by heavily utilizing robotic technology in combination with automated logistics and production lines.

Thomas Bock is a professor of building realization and robotics at Technische Universität München (TUM). His research has focussed for thirty-five years on automation and robotics in building construction, from the planning, prefabrication, on-site production, and utilization phases to the reorganization and deconstruction of a building. He is a member of several boards of directors of international associations and is a member of several international academies in Europe, the Americas, and Asia. He consulted for several international ministries and evaluates research projects for various international funding institutions. He holds honorary doctor and professorship degrees. Professor Bock serves on several editorial boards, heads various working commissions and groups of international research organizations, and authored or coauthored more than 400 articles.

Thomas Linner is a postdoctoral researcher in building realization and robotics and a research associate at TUM. He completed his dissertation (Dr.-Ing.) in 2013 in the field of automation and mass customization in construction with a particular focus on automated/robotic on-site factories. Dr. Linner is a specialist in the area of automated, robotic production of building "products" as well as in the conception and performance enhancement of those products through the embedding of advanced technology (service robots, microsystems technology). Today, more and more, issues related to innovation management are becoming key topics in his research. Dr. Linner is a frequently invited speaker at universities such as the University of Tokyo and Cambridge University.

Robotic Industrialization

AUTOMATION AND ROBOTIC TECHNOLOGIES FOR CUSTOMIZED COMPONENT, MODULE, AND BUILDING PREFABRICATION

Thomas Bock
Technische Universität München

Thomas Linner
Technische Universität München

CAMBRIDGE
UNIVERSITY PRESS

32 Avenue of the Americas, New York, NY 10013-2473, USA

Cambridge University Press is part of the University of Cambridge.

It furthers the University's mission by disseminating knowledge in the pursuit of education, learning, and research at the highest international levels of excellence.

www.cambridge.org
Information on this title: www.cambridge.org/9781107076396

First published 2015

A catalog record for this publication is available from the British Library.

Library of Congress Cataloging in Publication Data
Bock, Thomas, 1957–
Robotic industrialization : automation and robotic technologies for customized component, module, and building prefabrication / Thomas Bock, Technische Universität München, Thomas Linner, Technische Universität München.
   volumes cm
Includes bibliographical references and index.
ISBN 978-1-107-07639-6 (hardback : alk. paper)
1. Robots, Industrial.   I. Linner, Thomas, 1979–   II. Title.
TS191.8.B635   2014
629.8′92–dc23   2015022576

ISBN 978-1-107-07639-6 Hardback

. . . .

Legal Disclaimer

Contents

Acknowledgements

Construction automation gained momentum in the 1970s and 1980s in Japan, where the foundations for real-world application of automation in off-site building manufacturing, single-task construction robots, and automated construction sites were laid. This book series carries on a research direction and technological development established within this "environment" in the 1980s under the name *Robot-Oriented Design*, which was a focal point of the doctoral thesis of Thomas Bock at the University of Tokyo in 1989. In the context of this doctoral thesis many personal and professional relationships with inventors, researchers, and developers in the scientific and professional fields related to the construction automation field were built up. The doctoral thesis that was written by Thomas Linner (*Automated and Robotic Construction: Integrated Automated Construction Sites*) in 2013 took those approaches further and expanded the documentation of concepts and projects. Both theses form the backbone of the knowledge presented in this book series.

The authors would first like to express their deepest gratitude to Prof. Dr. Yositika Uchida, Prof. Dr. Y. Hasegawa, and Prof. Dr. Umetani (TIT); Dr. Tetsuji Yoshida, Dr. Junichi Maeda, Dr. Yamazaki, Dr. Matsumoto, Mr. Abe, and Dr. Ueno (Shimizu); Prof. Yashiro, Prof. Matsumura, Prof. Sakamura, Prof. Arai, Prof. Funakubo, Prof. Hatamura, Prof. Inoue, Prof. Tachi, Prof. Sato, Prof. Mitsuishi, Prof. Nakao, and Prof. Yoshikawa (UoT); Dr. Oiishi and Mr. Fujimura (MHI); Dr. Muro, Kanaiwa, Miyazaki, Tazawa, Yuasa, Dr. Sekiya, Dr. Hoshino, Mr. Arai, and Mr. Morita (Takenaka); Dr. Arai, Dr. Chae, Mr. Mashimo, and Mr. Mizutani (Kajima); Dr. Shiokawa, Dr. Hamada, Mr. Furuya, Mr. Ikeda, Mr. Wakizaka, Mr. Suzuki, Mr. Kondo, Mr. Okuda, Mr. Harada, Mr. Imai, Mr. Miki, and Mr. Doyama (Ohbayashi); Dr. Yoshitake, Mr. Sakou, Mr. Takasaki, Mr. Ishiguro, Mr. Morishima, Mr. Kato, and Mr. Arai (Fujita); Mr. Nakamura, Mr. Tanaka, and Mr. Sanno (Goyo/Penta Ocean); Mr. Nagao and Mr. Sone (Maeda); Mr. Ohmori, Mr. Maruyama, and Mr. Fukuzawa (Toda); Dr. Ger Maas (Royal BAM Group); Dr. Espling, Mr. Jonsson, Mr. Fritzon, Mr. Junkers, Mr. Andersson, and Mr. Karlson (Skanska); Mrs. Jansson, Mrs. Johanson, Mr. Salminen, Mr. Apleberger, Mr. Lindström, Mr. Engström, and Mr. Andreasson (NCC); Mr. Hirano (Nishimatsu); Mr. Weckenmann (Weckenmann); Mr. Ott and Mr. Weinmann (Handtech); Mr. Bauer and Mr. Flohr (Cadolto); Mr. Kakiwada, Mr. Ohmae, Mr. Sawa, Mr. Yoshimi, Mr. Shimizu, and Mr. Oki (Hazama Gumi); Mr. Masu, Mr. Oda, and Mr. Shuji (Lixil); Mr. Yanagihara,

Mr. Oda, Mr. Kanazashi, and Mr. Mitsunaga (Tokyu Kensetsu); Mr. Okada (Panasonic); Mr. Nakashima and Mr. Kobayashi (Kawasaki HI); Mr. Tazawa (Ishikawajima HI); Mr. Okaya (IHI Space); Dr. Morikawa and Mr. Shirasaka (Mitsubishi Denki); Mr. Maeda, Mr. Matsumoto, and Mr. Kimura (Hitachi Zosen); Mr. Shiroki and Mr. Yoshimura (Daiwa House); Mr. Hagihara and Mr. Hashimoto (Misawa Homes); Mr. Okubo (Mitsui Homes); Dr. Itoh, Mr. Shibata, Mr. Kasugai, Mr. Kato, and Mr. Komeyama (Toyota Homes); Mr. Hirano, Mr. Kawano, and Mr. Takahashi (Toyota Motors); Dr. K. Ohno (+); Dr. Misawa, Dr. Mori, Nozoe, Tamaki, Okada, and Dr. Fujii (Taisei); Prof. P. Sulzer and Prof. Helmut Weber (+); Prof. Dr. Fazlur Khan (+); Prof. Dr. Myron Goldsmith (+ IIT); Prof. A. Warzawski (+ Technion); Prof. R. Navon, Prof. Y. Rosenfeld, Prof. Dr. S. Isaac, Prof. A. Mita, Prof. A. Watanabe, Prof. Kodama, and Prof. Suematsu (Toyota National College of Technology); Mr. Suketomo, Mr. Yoshida, Mr. Oku, and Mr. Okubo (Komatsu); Dr. Ogawa and Mr. Tanaka (Yasukawa); Dr. Inaba (Fanuc); Mr. Usami (Hitachi RI); Mr. Ito and Mr. Hattori (Mitsubishi RI); Mr. Noda, Mr. Ogawa, and Mr. Yoshimi (Toshiba); Mr. Senba (Hitachi housetec); Mr. Sakurai, Mr. Hada, Mr. Sugiura, Mr. Yoshikawa, Mr. Shimizu, Mr. Tanaka, Mr. Nishiwaki, and Mr. Nomura (Sekisui Heim); Mr. Aoki, Mr. Watanabe, Mr. Kotani, and Mr. Kudo (Sekisui House); Mr. Takamoto, Mr. Mori, Mr. Baba, Mr. Sudoh, Mr. Fujita, and Mr. Hiyama (TMsuk); Prof. Dr. T. Fukuda, Prof. Iguchi, Prof. Ohbayashi, Prof. Tamaki, Mr. Yanai, and Mr. Hata (JRA); Mr. Uchida (JCMA); Dr. Kodama (Construction Ministry Japan); Dr. Sarata (AIST); Mr. Yamamoto (DoKen); Mr. Yamanouchi (KenKen); Mr. Yoshida (ACTEC); Prof. Dr. Wolfgang Bley, Prof. Dr. G. Kühn, and Jean Prouve (+); Dr. Sekiya, Dr. Yusuke Yamazaki, Prof. Dr. Ando, Prof. Dr. Seike, Prof. Dr. J. Skibniewski, Prof. Dr. C. Haas, Dr. R. Wing, Prof. Dr. David Gann, Prof. Dr. Puente, Prof. Dr. P. Coiffet, Prof. Dr. A. Bulgakow, Prof. Dr. Arai, Prof. Kai, Prof. Dr. A. Chauhan, Prof. Dr. Spath, Dr. M. Hägele, Dr. J.L. Salagnac, and Prof. Cai (HIT); Prof. Dr. Szymanski (+); Prof. Dr. F. Haller (+); Prof. Dr. Tamura and Prof. Dr. Han.

The authors would like to express their very sincere gratitude to the Japanese guest Professors Dr. K. Endo and Dr. H. Shimizu for supporting and advising the authors during their stays at the authors' chair in Munich. Furthermore, the authors are very thankful to all the companies in the field of component manufacturing, prefabrication, and on-site automation outlined throughout the book series for the information, analyses, pictures, and details of their cutting-edge systems and projects that they shared and provided.

The authors are indebted to the AUSMIP Consortium and the professors and institutions involved for supporting their research. In particular, the authors extend their thanks to Prof. Dr. S. Matsumura, Prof. Dr. T. Yashiro, Prof. Dr. S. Murakami, Dr. S. Chae, Prof. Dr. S. Kikuchi, Assistant Prof. Dr. K. Shibata, Prof. Dr. S.-W. Kwon, Prof. Dr. M.-Y. Cho, Prof. Dr. D.-H. Hong, Prof. Dr. H.-H. Cho, Prof. A. Deguchi, and Prof. B. Peeters and their students and assistants for supporting the authors with information and organizing and enabling a multitude of important site visits. Furthermore, the authors gratefully acknowledge the International Association for Automation and Robotics in Construction whose yearly conferences, activities, publications, and network provided a fruitful ground for their research and a motivation to follow their research direction.

The authors are indebted in particular to W. Pan, K. Iturralde, B. Georgescu, and M. Helal for their great support in the phase of completing the volume series. The authors thank T. Isobe, S. Kitamura, and Prof. T. Seike for their support and Dr.-Ing. C. Georgoulas, A. Bittner, J. Güttler, and S. Lee for their advice and support. In addition, the authors thank A. Liu Cheng, B. Toornstra, I. Arshad, P. Anderson, N. Agrafiotis, K. Pilatsiou, G. Anschau Rick, and F. Gonzalez.

The authors are sincerely grateful also to their numerous motivated students within the master course Advanced Construction and Building Technology who have, as part of lab or coursework, contributed to the building of the knowledge presented and who have motivated the authors to complete this book series.

Glossary

Alignment and accuracy measurement system (AAMS): An AAMS creates a feedback loop between a system that measures how accurately components are positioned and an alignment system (e.g., a motorized unit attached temporarily to the joint of a column component) that automatically moves or aligns the *component* into the desired final position.

Assembly: In this book, assembly refers to the production of higher-level components or final products out of *parts* and lower-level *components*. The process of assembly of an individual *part* or *component* to a larger system involves positioning, alignment, and fixation operations. *Upstream* processes dealing with the generation of elements for assembly are referred to as *production*.

Automated guided vehicle (AGV): Computer-controlled automatic or robotic mobile transport or logistics vehicle.

Automated/robotic on-site factory: *Structured environment* (factory or factory-like) setup at the place of *construction*, allowing *production* and *assembly* operations to be executed in a highly systemized manner by, or through, the use of machines, automation, and robot technology.

Batch size: The amount of identical or similar products produced without interruption before the manufacturing system is substantially changed to produce another product. Generally speaking, low batch sizes are related to high fixed costs and high batch sizes are related to low fixed costs.

Building component manufacturing (BCM): BCM refers to the transformation of *parts* and low-level *components* into higher-level *components* by highly mechanized, automated, or robot-supported industrial settings.

Building integrated manufacturing technology: Automation technology, microsystem technology, sensor systems, or robot technology can be directly integrated into buildings, units, or components as a permanent system. Technology used to manufacture the building can thus become a part of the building technology.

Bundesministerium für Bildung und Forschung (BMBF): Federal Ministry of Education and Research in Germany. The BMBF funds education, research, and technical development in a multitude of industrial fields.

Capital intensity: The capital intensity (also referred to as workplace cost) is calculated by dividing the capital stock (assets, devices, and equipment used to transform/manufacture the outcome) by the number of employees in the industry. The construction industry not only has one of the lowest capital stocks, but also capital intensity is by far the lowest compared to other industries in Germany.

Chain-like organization: In a chain-like organization, the *flow of material* between individual workstations is highly organized and fixed, and a material transport system linking the stations exists.

Climbing system (CS): Automated/robotic on-site factories require, especially in the manufacturing of vertically oriented buildings, a system that allows the *sky factory* (SF) to rise to the next floor, once a floor level has been completed. Most SFs, therefore, rest on stilts that transfer the loads of the SF to the building's bearing structure, or to the ground. Other CSs are able to climb along a central core, pushing up the building or, in the case of the manufacture of horizontally orientated buildings, for example, to enable the factory to move horizontally. In some cases, in addition to climbing, CSs are used to provide a fixture or template for the positioning of components by *manipulators*. Because of the enormous forces necessary to lift SFs, hydraulic systems and screw presses are common actuation systems.

Closed loop resource circulation: Systems for avoiding waste and reduction of resource consumption, by integrating concepts such as reverse logistics, remanufacturing, and recycling. Material or product that flows on a factory, utilization, and deconstruction level can be related back to the manufacturing system to close the loop.

Closed sky factory (CSF): *Sky factory* that completely covers and protects the workspace in *automated/robotic on-site factories*, thus allowing the installation of a fully *structured environment* that erases the influence of parameters that cannot be 100% specified (e.g., rain, wind).

Component: In this book, in a hierarchical modular structure, components can be divided into lower-level components and higher-level components. Components consist of subelements of *parts* and lower-level components. Higher-level components can be assembled into *modules* and *units*.

Component carriers: Component carriers and pallets (special types of component carriers) play an important role in logistics. In many cases, parts, components, or final products cannot be directly handled or manipulated by the logistics system. Component carriers and pallets act as mediators between the handled material and the actual logistics system.

Computer-aided design/computer-aided manufacturing (CAD/CAM): From the 1980s on, the novel and highly interdisciplinary research and application field CAD/CAM was formed, which aimed at integrating computerized tools and systems from the planning and engineering field with manufacturing and machine control systems to allow for a more-or-less direct use of the digital design data for automated and flexible manufacturing. The field evolved further toward *computer-integrated manufacturing*.

Computer-aided quality management (CAQM): Control of quality by software, made possible through the linking of manufacturing systems with computer systems.

Computer-integrated manufacturing (CIM): From the 1990s on, the combined *computer-aided design/computer-aided manufacturing* approach evolved into the CIM approach. The focus was then broader and the idea was that more and more fields and tools, and also business economic issues (e.g., computer-aided forecasting or demand planning), could be integrated by computerized systems to form continuous process and information chains in manufacturing that span all value-adding nodes in the value system.

Connector system: The development of connector systems that connect complex *components* in a robust way to each other is a key element in complex products such as cars, aircraft, and buildings in particular. To support efficient *assembly*, connector systems can, for example, be compliant or plug-and-play–like. Connector systems can also be designed to support efficient disassembly, remanufacturing, or recycling.

Construction: Activities necessary to build a building on-site. Construction, in this book, is interpreted as being a manufacturing process, and accordingly buildings are seen as "products."

Cycle Time: Important on the workstation level: The cycle time refers to the time allowed for all value-adding activities performed by humans and machines at a workstation within a network of workstations.

Degree of freedom (DOF): In a serial kinematic system each joint gives the systems, in terms of motion, a DOF. At the same time, the type of joint restricts the motion to a *rotation* around a defined axis or a *translation* along a defined axis.

Depth of added value: The depth of added value (e.g., measured as a percentage of the total cost of the product) refers to the total amount of value-adding activities, and thus in general to the amount of value-adding steps, realized by the *original equipment manufacturer* (OEM) or final integrator. A high depth of added value means that a large number of value-adding activities are being realized by the OEM (e.g., Henry Ford). A low depth of added value means that a low number of value-adding activities is being realized by the OEM (e.g., Dell, Smart).

Design for X (DfX): DfX strategies aim at influencing design-relevant parameters to support production, assembly maintenance, disassembly, recycling, and many other aspects related to a product's life cycle. In this book, DfX strategies are classified into four categories: DfX related to *production/assembly*, to product function, to product end-of-life issues, and to business models. In this book, *robot-oriented design* is seen as an augmentation or extension of conventional DfX strategies, consequently aiming more at the efficient use of automation and robotic technology in all four categories.

Deutscher Kraftfahrzeug-Überwachungs-Verein (DEKRA): Major German consultant and surveyor association that evaluates technical artefacts, such as cars and buildings, and defines quality and the causes of defects.

Efficiency: Efficiency can be defined as the relationship between an achieved result and the combination of factors of production. Whereas *productivity* expresses an input-to-output ratio, with a focus on a single factor of production, efficiency considers multiple factors and their combination and interrelation. Productivity can be an indication of efficiency, and efficiency itself for economic feasibility.

End-effector: The element of machines, automation systems, or robot technology that makes contact with the object to be manipulated in *manufacturing* is called the end-effector. In most cases end-effectors are modularly separable from the base system. End-effectors have a certain degree of *inbuilt flexibility*.

Factory external logistics (FEL): FEL refer to logistics systems that connect the supply network to the factory integrating and assembling the supplied *parts*, *components*, *modules*, or *units*. FEL influences the organization of the manufacturing system, *factory internal logistics*, and the factory layout.

Factory internal logistics (FIL): FIL refers to logistics systems that manipulate *parts*, *components*, *modules*, *units*, or the finished product within a manufacturing setup or factory, for example, for the transportation between various stations. Other examples include mobile and non-rail–guided transport systems, overhead crane-type material transportation systems, fixed conveyor systems allowing a component carrier or the product itself to travel in a horizontal direction in fixed lanes, and fixed conveyor systems allowing a component carrier or the product itself to travel in a vertical direction in fixed lanes. Novel cellular logistics robots combine capabilities of unrestricted mobility with horizontal and vertical transport capabilities and can travel freely and self-organize with other systems.

Factory roof structure: Structure that allows the workspace on the construction site to be covered (and therefore to be protected from outside influences such as wind, rain, or sun) and thus creates the basis for a *structured environment*. Often used as a platform for the attachment of other subsystems, such as a *climbing system*, *horizontal delivery system*, and *overhead manipulators*.

Final integrator: In this book, a final integrator refers to an entity in a value chain or value system that integrates major components into the final product. Within the OEM model, the final integrator is called *original equipment manufacturer*.

Fixed-site manufacturing: *Off-site manufacturing* or *on-site manufacturing* system that stays at a fixed place during final *assembly*.

Floor erection cycle (FEC): Time necessary to erect and finish (including technical installations and general interior finishing) a standard floor with an *automated/robotic on-site factory*.

Flow-line organization: In a flow-line organization, individual workstations do not have a fixed *flow of material*, but a general directional *flow of material* (e.g., within a factory segment or a factory).

Flow of material: Refers to material and product streams in relation to space and time that take place during the completion of a specific product in a manufacturing system and the supply network connected to it. The *efficiency* of the flow of material is determined by the arrangement of equipment, the factory layout, and the logistics processes.

Frame and infill (F&I): F&I strategies are used in a variety of industries, including the aircraft, automotive, and building industries. The idea of an F&I strategy is to use a bearing frame structure as a base element that is subsequently equipped with *parts*, *components*, systems, *modules*, and so forth during the manufacturing process. The frame thus functions as a *component carrier*. In the aircraft industry, the fuselage is interpreted as such a frame; in the automotive industry it is the car body or chassis; and in the building industry it can be seen, for example, in the form of two-dimensional (e.g., Sekisui House) or three-dimensional steel frames (Sekisui Heim).

Ground factory (GF): *Structured environment* (factory or factory-like) setup on the construction site on the ground level of the building as part of an *automated/robotic on-site factory*.

Group-like organization: In a group-like organization, individual workstations are bound together in groups. Those groups can refer to workstations with similar means of production or to workstations with complementary *means of production*. The *flow of material* between those groups can be either fixed or flexible.

Horizontal delivery system (HDS): System that transports, positions, and/or assembles *parts/components* on the construction site on a floor level.

Idle time: The unproductive standstill of a machine from end of completion to the beginning of the processing of the next material. Bottleneck operations, for example, may – when workstations are directly connected without a buffer – lead to material having to wait for a certain time until the next material can be processed and to an unproductive standstill of other workstations that are faster in processing the material.

Inbuilt flexibility: The changes in a manufacturing system can be realized without major physical or modularization enabled changes (e.g., exchange of systems, workstations, robots, *end-effectors*), but by reprogramming the existing system instead. A standard robot with 6 *degrees of freedom* (6-DOF robot) with an end-effector for welding, for example, has a high degree of flexibility and can be reprogrammed for a huge variety of welding operations within a given workspace.

Joint of a manipulator: A *manipulator* consists of at least one kinematic pair consisting of two rigid bodies (links) interconnected with a joint. The following types of joints can be distinguished: revolute joint, prismatic joint, and spherical joint.

Just in sequence (JIS): Various *parts*, *components*, and *products* are delivered from *upstream* to *downstream* workstations in the sequence in which they are handled or processed when they reach the *downstream* work stations. JIS can be performed internally within a factory or in relation to a supplier of an *original equipment manufacturer* (OEM). JIS is in most cases closely connected to *just in time* (JIT).

Just in time (JIT): Stocks and buffers are eliminated, and *parts*, *components*, and products are delivered from *upstream* to *downstream* workstations at the right time and at the right quantity. JIT can be performed internally within a factory or in relation to a supplier of an *original equipment manufacturer*. JIT is in most cases closely connected to *just in sequence*.

Kinematic base body: The combination of links and joints forms kinematic bodies that allow basic manipulation operations within a geometrically definable work space (e.g., Cartesian *manipulator*, gantry *manipulator*, cylindrical *manipulator*, spherical *manipulator*). Those kinematic base bodies consider mainly the first three axes, and thus refer mainly to *positioning* activity. For *orientation*, further *degrees of freedom* and kinematic combinations can be added on top of those base bodies.

Kinematics: Kinematics focuses on the study of geometry and motion of automated and robotic systems. It describes parameters such as position, velocity, and acceleration of joints, links, and *tool centre points* to generate mathematical models creating the basis for controlling the actuators and for finding optimized trajectories for the motions of the system. *Manipulators* are a kinematic system consisting of a multitude of kinematic subsystems, of which the kinematic pair is the most basic entity.

Large-scale prefabrication (LSP): *Off-site manufacturing* of high-level *components*, *modules*, or *units* in very large quantity by a production-line–based, automation and robotics-driven factory or factory network, interconnected in an *OEM-like integration structure*.

Link of a manipulator: A *manipulator* consists of at least one kinematic pair, consisting of two rigid bodies (links) interconnected with a *joint*.

Logistics systems: Logistics can be defined as the transport of material within manufacturing systems and supply networks. Logistics is a kind of manipulation of an object by humans, tools, machines, automation systems, and robots (or combinations of those), positioning and orientating objects to be transported or processed in a three-dimensional space. However, logistics operations do not change or transform the material directly. Logistics systems can be characterized according to various scales, such as assembly system scale, factory internal scale (*factory internal logistics*), factory external scale (*factory external logistics*).

Manipulator: In this book series, a manipulator refers to a system of multiple links and joints that performs a kinematic motion. Depending on the ratio of autonomy and intelligence, manipulators can be machines, automated systems, or robots.

Manufacturing: In this book, manufacturing refers to systems that produce products. Manufacturing integrates *production* (parts or low-level component production) and *assembly* processes.

Manufacturing lead time: Time necessary to complete a product within a given manufacturing system, factory, or factory network.

Mass customization (MC): MC strategies combine advantages of *workshop-like* and *production-line*–like manufacturing, and thus product differentiation–related competitive advantages, with mass-production–like efficiency. On the product side, MC demands that a product combines customized and standardized elements, for example, through *modularity*, *platform strategies*, and *frame and infill* strategies to be able to efficiently produce it in an industrialized manner. On the manufacturing side, MC demands highly flexible machines, automation systems, or robot technology that removes the need for human labour in the customization process.

Material handling, sorting, and processing yard (MHSPY): Subsystem of an *automated/robotic on-site factory*; often related to the *ground factory*. An MHSPY can be a covered environment and/or can be equipped with *overhead manipulators* and allows the simplification or automation of the picking up of *components* from delivering *factory external logistics* in a *just in time* and *just in sequence* manner. An MHSPY can also be used to transform *parts* and low-level *components* into higher-level components on-site. In *automated/robotic on-site factories* used to deconstruct buildings, MHSPYs can be used to transform higher-level *components* into lower-level *components* and *parts*.

Means of production: Means of production can be classified into human resources, equipment, and material to be transformed.

Modular flexibility: When the change of a product or the variation of a product is so intense that the *inbuilt flexibility* of a manufacturing system, a machine, or an *end-effector* cannot cope with it, a rearrangement or extension of the manufacturing system on the basis of *modularity* becomes necessary. *Modularity* can be generic (predefined process or system modules) or unforeseen (use/design of completely new modules, new configurations).

Modularity: Modularity refers to the decomposition of a structure or system into rather independent subentities. It can cover the functional realm as well as the physical realm. If structures or systems are nearly impossible to decompose, on both functional and physical levels, the artefacts are referred to as "integral." If systems can clearly be decomposed, on both functional and physical levels, artefacts are referred to as "modular." Clear modularity is, in construction practice, still a rare phenomenon, and conventional buildings show basic characteristics of integral product structures. *Automated/robotic on-site factories*, however, require strict modularity.

Module: In this book, in a hierarchical modular structure, modules represent elements on a hierarchical level above high-level *components*. *Parts*, *components*, and modules can be assembled into *units*, which are ranked higher than modules.

nth, $n-1$, $n-2$, $n-X$ floors: Inside the main factory (e.g., a *sky factory*) of *automated/robotic on-site factories*, work (component installation, welding, interior finishing, etc.) is done in parallel on several floors (*n*-floors). The *n*th floor represents the uppermost floor in which work takes place in parallel and the $n-X$ floors represent the floors below this floor in which work takes place in parallel.

OEM-like integration structure: Value systems or *parts/components* integration structures that do not fully follow the *OEM model* but show characteristics of it.

OEM model: An *original equipment manufacturer* relies on suppliers, which, according to their rank in the supply chain, are called Tier-*n* suppliers. The model explains the general *flow of material* as well as the flow of information during development of the product and its subcomponents.

Off-site manufacturing (OFM): *Components* or complete products are manufactured in a *structured environment* distant from the final location where they are finally used. *Components* or complete products can be packed and shipped or are mobile (e.g., car, aircraft).

One-piece-flow (OPF): OPF refers to a highly systemized and *production-line–* based manufacturing system in which each *component* or product assembled can be different.

On-site manufacturing (ONM): Products such as buildings, towers, and bridges have to be produced on site by ONM systems at the location at which they are to be finally used as they cannot be moved or shipped as an entity.

Open sky factory (OSF): *Sky factory* covering and protecting the workspace in *automated/robotic on-site factories*, and, in contrast to *closed sky factories* allows only the installation of a partly *structured environment* that at least (compared to conventional construction) minimizes the influence of parameters that cannot be 100% specified/foreseen (e.g., rain, wind).

Original equipment manufacturer (OEM): Integrates and assembles *components* and subsystems coming from sub-factories and suppliers to the final product within the *OEM model*. Companies or entities in the value chain that do not fully follow the *OEM model* but show characteristics of it are also referred to as *final integrators*.

Overhead manipulators (OMs): OMs operate within off- or on-site *structured environments* and in *automated/robotic on-site factories* are often the central elements of the *horizontal delivery system*. On the one hand, OMs (e.g., gantry-type OMs) allow the precise manipulation of components of extreme weights and at high speed, which cannot, for example, be accomplished by conventional industrial robots such as anthropomorphic manipulators. On the other hand, OMs require a simplification of the assembly process by *robot-oriented design*, as their workspace and their ability to conduct complex positioning and orientation tasks are limited.

Part: In this book, in a hierarchical modular structure, parts represent elements on a hierarchical level below *components*.

Performance multiplication effect (PME): Once significant productivity increases in an industry can be achieved (i.e., by switching from crafts-based to machine-based manufacturing), an upward spiral starts: high productivity can become a driver of the financing elements for innovations related to even better machines, processes, and products and thus even higher productivity. This phenomenon was/can be observed in many non-construction industries (e.g., textile industry, automotive industry, shipbuilding) and is in this book series referred to as the PME.

Platform strategy: A platform is a basic framework; a set of standards, procedures, or parts; or a basic structure that contains core functions of a product. A platform allows for the highly efficient production of customized products, as it allows for the platform to be mass-produced and to wear individual modules on top of it, which can be customized or personalized.

Positioning and orientation: For unrestricted positioning of an object within a defined space, or within x, y, and z coordinates, at least 3 *degrees of freedom* are necessary (also referred to as forward/back, left/right, up/down). For unrestricted orientation of an object around a *tool centre point*, at least 3 *degrees of freedom* are necessary (also referred to as yaw, pitch, and roll).

Production: In this book, production refers to the generation of basic *parts* or low-level *components*. It includes transformation of raw material into *parts*. *Downstream* processes dealing with the joining of elements generated within production are referred to as *assembly*. *Manufacturing* includes production and *assembly* processes.

Production line–like organization: In a production line–like organization, the flow of material between individual workstations is fixed; a material transport system links the stations and the *cycle times* of the individual workstations are synchronized.

Productivity: Productivity = Output (quantity)/Input (quantity). Productivity quantitatively expresses an input-to-output ratio, with a focus on a single (input) means of production or a single (input) factor of production. Productivity indices concerning the type of factor are, for example, work, capital, material, resource, and machine productivity.

Pulling production: Refers to a production system in which products are manufactured only on the basis of actual demand or orders. *Parts*, *components*, and products required are pulled from *upstream*, according to the actual demand. It might refer to the whole manufacturing system, as well as to individual workstations or groups of workstations. Examples: *Toyota Production System*, Sekisui Heim, Toyota Home.

Pushing production: Refers to the continuous production of elements/products in a certain fixed amount based on predictions or assumptions. Without taking into consideration the actual demand in *downstream* process steps, *parts*, *components*, and products are pushed through individual stations. It might refer to the whole manufacturing system or to individual workstations or groups of workstations. Example: Henry Ford's mass production.

Radio frequency identification tag (RFID): RFID tags are inexpensive tags that can be attached to *components*, *modules*, *units*, or products. RFID readers can be integrated into floors or placed over gates and can then identify the object passing by. They can be distinguished between simple low-cost passive tags and more complex active tags. Advanced readers can read multiple tags at once.

Real-time economy (RTE): Macroeconomic view of the impact of the multitude of changes our economy, manufacturing technology, and the relation between customers and businesses undergo. It targets the fulfilment of customer demands and requests in near real time. Products and services are processed within a few hours and delivered within a few days.

Real-time monitoring and management system (RTMMS): Data from sensor systems, as well as from the servomotors/encoders of the *vertical delivery system* (VDS) and *horizontal delivery system* (HDS), along with information obtained from cameras monitoring all activities (including human activities) in *automated/robotic on-site factories* are used to create a real-time representation of equipment activity and of construction progress. Furthermore, barcode systems often allow the representation and optimization of the *flow of material*, allowing equipment (such as VDS, HDS, or OM) to identify the *component* being processed. In most cases, real-time monitoring and management is done in a fully computerized on-site control centre. An RTMMS simplifies progress and quality control and reduces management complexity.

Re-customization: Remanufacturing strategy that allows a building to be disassembled and for major *components* or *units* to be refurbished and equipped with new *parts* or *modules* on the basis of mechanized or automated manufacturing systems to meet changed or new (individual) customer demands.

Robot-oriented design (ROD): ROD is concerned with the co-adaptation of construction products and automated or robotic technology, so that the use of such technology becomes applicable, simpler, or more efficient. The concept of ROD was first introduced in 1988 in Japan by T. Bock and served later as the basis for automated construction and other robot-based applications.

Rotation: A term used to describe a kinematic structure. A revolute joint allows an element of a machine or manipulator to rotate around an axis and in a serial kinematic system adds 1 *degree of freedom* to the system.

Selective compliance articulated robot arm (SCARA): Developed by Yamanashi University in Japan in the 1970s. It combines two revolute and one prismatic joint so that all motion axes are parallel. This configuration and the thus enabled allocation of the actuators are advantageous for the stiffness, repeatability, and speed with which the robot can work. Owing to its simplicity, the SCARA is also a relatively cheap robot system. It laid the foundation for the efficient and cheap production, and thus the success of, for example, Sony's Walkman.

Single-task construction robot (STCR): STCRs are systems that support workers in executing one specific construction process or task (such as digging, concrete levelling, concrete smoothening, and painting) or take over the physical activity of human workers that would be necessary to perform one process or task.

Sky factory (SF): Structured environment (factory or factory-like) setup on the construction site as part of *automated/robotic on-site factories*. SFs cover the area where building *parts* and *components* are joined to the final product and rise vertically with the upper floor of a building through a *climbing system*. SFs can enclose and protect the work environment completely (*closed sky factory*) or only partly (*open sky factory*).

Slip forming technology: Moving or self-moving form that allows casting concrete structures such as columns, walls, or towers on site in a systemized manner on the construction site.

Stilts: In this book, stilts refer to elements of automated/robotic on-site factories. The *sky factories* of *automated/robotic on-site factories* often use stilts (made of steel) integrated within the *climbing system* to be rested on the building that they are manufacturing. Stilts can be lifted and lowered via the *climbing system*, thus allowing the *sky factory* to move on top of the building's steel column structure.

Structured environment (SE): In factories or factory-like environments, work tasks, workspaces, assembly directions, and many other parameters (e.g., climate, light, temperature) can be standardized and precisely controlled. The structuring of an environment creates the basis for the efficient use of machines, automation, and robot technology. The structuring of an environment includes the protection from uncontrollable factors such as wind, rain, sun, and non-standardized human work activity.

Superstructure: The concept of dividing a building into superstructure and substructure is an approach that introduces the concept of hierarchies to a building's structure and *components* and thus can serve as a basis for possible *modularity* in construction. Goldsmith introduced the idea of making the transmission of forces within high-rise buildings by the superstructure to a clearly visible architectural element in his thesis (1953). A superstructure can serve as a platform or frame that allows customization by further infill and is thus closely connected to *frame and infill* strategies.

Supply chain: The supply chain connects value-added steps and transformational processes across the border of individual factories or companies. Its aim is to interconnect all processes and workstations to complete a product informationally and physically, in order to create uninterrupted on-demand or in-stock *flow of material*.

Sustainability in manufacturing: Manufacturing systems can be designed to be efficient and to equally meet economic, environmental, and social demands. In this book, sustainability in manufacturing refers predominantly to the ability of a manufacturing system to reduce consumption of resources and the generation of waste.

Technology diffusion: Technology diffusion describes the step-by-step spread of a technology throughout industry or as, for example, computer technology throughout society. To simplify the adoption of novel technologies and increase their application scope over time, novel technologies have to be made less expensive and less complex and be split into individual modular elements. Technology diffusion therefore often is accompanied by a switch from centralized to rather decentralized applications of the novel technology.

Tier-*n* supplier: Suppliers are, according to their rank in the supply chain, called Tier-*n* suppliers. A Tier-1 supplier is a first-rank supplier that relies on *components* from a Tier-2 supplier; a Tier-2 supplier relies on *components* from Tier-3 suppliers; and so on.

Tool centre point (TCP): The *end-effector* is a tool that is carried by the kinematic system. For each end-effector, a tool TCP is defined as the reference point for kinematic calculations.

Toyota Production System (TPS): The TPS is a logical and consequent advancement of the concept of mass production to a more flexible and adaptive form of demand-oriented manufacturing, developed by Toyota between the 1960s and 1970s. From the 1980s, TPS principles gained worldwide recognition and today they form the conceptual basis for manufacturing systems around the world. Concepts such as *just in time*, Kaizen, *Kanban, pulling production*, failure-free production, and *one-piece flow* have their origins in the TPS.

Transformational process: Any organization and its manufacturing system transform inputs (information, material) into outputs (products, services). The transformation is performed by the organization's structural setting and its means of production, resulting in a specific combination and interaction of workers, machines, material, and information.

Translation: A term used to describe a kinematic structure. A prismatic joint allows an element of a machine or manipulator to move in a given trajectory along an axis and, in a serial kinematic system, adds 1 *degree of freedom* to the system.

Tunnel boring machine (TBM): TBMs mechanize and automate repetitive processes in tunnel construction. TBMs are self-moving underground factories that more or less automatically perform excavation, removal of excavated material, and supply and positioning of precast concrete segments. TBMs are equal in many ways to automated construction sites (and the production of a building's main structure by *automated/robotic on-site factories*).

24/7-mode: The operation of a factory, processes, or equipment without major interruption 24 hours a day, 7 days a week. Requires the work environment to be structured (*structured environments*) and protected from influencing factors such as weather and the day/night switchover.

Unit: In this book, units refer to high-level building blocks. *Parts*, *components*, and *modules* can be assembled into units, which are higher ranked than *modules*. Units are completely finished and large three-dimensional building sections, manufactured off site.

Unit method: Sekisui Heim, Toyota Home, and Misawa Homes (Hybrid) break down a building into three-dimensional *units*. Those *units* are realized on the basis of a three-dimensional steel space frame, which, on the one hand is the bearing (steel) structure of the building, and on the other hand can be placed on a production line where it can be almost fully equipped with technical installations, finishing, kitchens, bathrooms (plumbing *units*), and appliances.

Upstream/downstream: Refer in this book to processes or activities in a value chain or manufacturing chain that are conducted before (upstream) or after (downstream) a certain point.

Urban mining: Refers to strategies that allow the city and especially its building stock to be a "mine" for resources, *parts*, and *components*. Systemized deconstruction of buildings under controlled and structured conditions, as in *automated/robotic on-site factories*, are enablers of urban mining.

Vertical delivery system (VDS): System that transports *parts/components* on the construction site from the ground (e.g., material handling yard) to the floor level where the components are to be assembled. VDSs play an important role in most *automated/robotic on-site factories*.

Workbench-like organization: The product stays at a fixed station in the factory where it is produced or assembled manually or automatically through the use of various tools. The means of production are organized around this one station.

Workshop-like organization: In a workshop-like organization, the product and/or its components flow between workstations. The sequence is not fixed and the times products stay at a certain workstation vary with the product.

"Zero" waste factory: A factory that minimizes resource input and waste output and allows (almost) all generated waste to be recycled.

1 Introduction

In this volume, concepts, technologies, and developments in the field of building component manufacturing (BCM, outlined in Chapter 2) based on concrete, brickwork, wood, and steel materials, as well as building module manufacturing (outlined in Chapter 3) and large-scale prefabrication (LSP, outlined in Chapters 4 and 5) with the potential to deliver complex components and products, are introduced and discussed. BCM refers to the transformation of materials, parts and low-level components into higher level components through the use of highly mechanized, automated, or robot-supported industrial settings. The definitions of components share a common element; they are, more or less, a complex combination of individual preexisting basic elements, parts and/or lower level components. BCM should also be distinguished from the manufacturing of more complex modules (e.g., prefabricated bath modules) or units (products of LSP companies, e.g., prefabricated three-dimensional building sections as manufactured, for example, by Toyota Home and Sekisui Heim).

For highly automated LSP, according to the original equipment manufacturer (OEM; see also Section 1.1) model, component manufacturers (BCM companies) represent Tier-1 or Tier-2 suppliers. Tier-1 suppliers deliver components directly to LSP companies such as Sekisui Heim, whereas Tier-2 suppliers would, for example, provide them to the suppliers of the bath or kitchen modules (building module manufacturers). For automated construction sites utilizing singe-task construction robots (STCRs, see **Volume 3**) or automated/robotic on-site factories (see **Volume 4**), low- and high-level components manufacturers (BCM, manufacturers of modules, LSP) again represent Tier-n suppliers.

In automotive manufacturing, for example, the Smartville factory (**Volume 1, Section 4.3.4**) demonstrates that the delivery of well prefabricated, high-level components to the final integrator and assembly line considerably reduces task variability, the amount of necessary assembly operations, organizational complexity and lead times and increases significantly the possibility to automate. Well-designed basic elements/parts/components are able to foster the creation of a structured environment (SE) in the receiving value added step. Therefore, as outlined in **Volume 1, Section 6.3**, in automated/robotic construction the whole value chain has to be considered, as each value added step holds the potential for prestructuring and simplification of processes (major success factor for efficient automation/use of robotic technology)

1

for the subsequent value-added steps. So, for example, the process of transformation of raw materials into basic elements by additive manufacturing holds the potential to create basic elements that are directly customized/optimized for the automated processing by a certain machine into complex parts or low-level components in a subsequent value added step. The manufacturing of parts or low-level components then can (e.g., through the embedding/imprinting of compliant joints, guiding elements, and sensors/tags for guiding the end-effectors picking them; see also **Volume 1**, **Section 6.5**) can simplify/foster assembly by a robot system into more complex components as, for example, walls or panels or units. Similarly, the module manufacturing and LSP industries can through the delivery of manufacturing optimized modules, panels and units reduce on-site assembly complexity (amount and variety of tasks) and thus foster the efficient use of robots operated on partly automated construction sites (SCTRs) or within highly automated/robotic on-site factories. Furthermore, BCM, module manufacturing and LSP are able to insert functionality in components (e.g., sensor elements, microsystem technology) that is able to foster features related to ambient robotics (**Volume 5**) as, for example, robot-enabled maintenance, recustomization, and other building integrated life-support services.

1.1 OEM Model and Manufacturing Strategy

In industries in which highly complex products are manufactured (automotive, aircraft, and in particular, building industry; see also **Volume 1, Chapter 5**) individual components are often so complex that a supplier must rely on other suppliers, thus leading to the OEM model (Figure 1.1; for more details, see **Volume 1**).

The concept of prefabrication, to which the aforementioned concepts belong, becomes increasingly important in our industrialized economies. In recent years, there has been an increase in the use of prefabrication, not only in building construction, but also in other industries such as automotive manufacturing, engine construction, and food supply. Time plays a big role in today's society and is a factor in many areas of various markets. The goal of prefabrication should be to improve the efficiency and performance of a product. The term efficiency (see also **Volume 1**, **Section 4.1**) encompasses many aspects, as the goal is not limited to pure cost reduction, but more so to the upholding of quality while saving time through the shortening of building phases, and reducing failure cost. The money saved is then available for the end-user, system operator, contractor or machine supplier – for example, to be reinvested into research towards superior product performance and thus to trigger a performance multiplication effect (PME; a basic concept in automated/robotic construction, discussed throughout **Volume 1**).

A well-planned manufacturing strategy is the key to successful prefabrication. A manufacturing strategy can be classified into hard and soft items. Hard items comprise decisions such as production capacity, factory network, selection of production technology, and vertical integration. Conversely, personnel/labour management, supplier management, production plan control, costing, and general management can be classified as soft items. The materials used determine both hard and soft production items. Brickwork, steel, concrete, carbon fibre composites, wood – every construction material that determines the primary and secondary structure of a building – has specific requirements and potentials. In addition, depending on

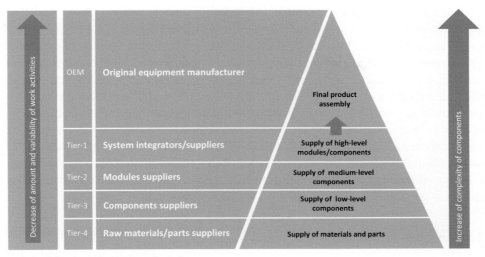

OEM model in general manufacturing industry

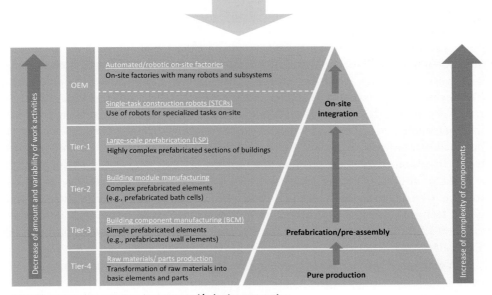

OEM-like integration structure in automated/robotic construction

Figure 1.1. OEM model. OEMs rely on suppliers, called Tier-*n* suppliers, according to their rank in the supply chain. The model explains the general flow of material as well as the flow of information during manufacturing of the product and its subcomponents (Authors' interpretation based on Kurek, 2004). In an OEM-like integration structure, well-designed building components are able to foster the creation of a structured environment (SE) in the receiving value added step. Therefore in automated/robotic construction the whole value chain has to be considered (from raw material transformation to final assembly and operation/deconstruction of buildings), as each value added step holds the potential for prestructuring and simplification of processes (major success factors for efficient automation/use of robotic technology) for the next value added step.

the material and local area, various construction types have been developed that must be synchronized with a specific manufacturing strategy. BCM and LSP refers to the transformation of materials, parts and low-level components into higher-level components through highly mechanized, automated, or robot-supported industrial settings and manufacturing systems. The term "manufacturing" is used in this volume (as well as in the other volumes in the series) as an umbrella term that covers both production (a process or set of processes that transform raw material into treatable basic elements, parts or low-level components) and assembly (a process or set of processes that joins various basic elements, parts and low-level components into medium/high-level components). Thus, BCM covers the transformation or preparation of individual materials and parts as well as their combination with other elements in a factory or factory-like, structured environments.

1.2 Analysis Framework

The analysis framework used in this volume was set up to enable identification and analysis of the relationship and interplay of products (and product performance), manufacturing strategies, manufacturing layouts and manufacturing technologies related to the factory-based off-site manufacturing of buildings. The previously identified (see **Volume 1, Chapter 4**) and discussed relevant topics in the areas of modularity, technology, and organization in manufacturing, automation, and robot technology form the basis for the analysis and outline in this volume. Furthermore, the possibility of supporting automated and robotic processes through robot-oriented design (ROD, see also **Volume 1, Chapter 6**) and generating customized/personalized (or better, industrially and robotically customized/personalized) products by industrialized and highly automated manufacturing systems is analysed. In addition, the analysed manufacturing strategies are related to the greater context of the construction industry (current situation, market shares, and history) as well as emerging topics (e.g., end-of-life strategies) and innovations. Concepts, technologies, and developments in the field of BCM, module manufacturing, and LSP described in this volume are analysed by the framework outlined in Table 1.1.

In general, it can be said that both wood and steel off-site manufacturing methods allow for generating higher level components, and thus higher added value when compared with brickwork and concrete off-site component manufacturing. As the large-scale off-site component manufacturing on basis of steel panels (e.g., Sekisui House) or three-dimensional steel units (e.g., Sekisui Heim) shows, steel structures allow for the generation of a carrier element, carrier frame or template that can be equipped in production line–like, automated SEs with various other parts and components. Furthermore, steel is a material that is easily processed (because of its weight and density) with high precision in an off-site SE. The use of automated systems, robots, and end-effector tools for the processing of steel in SEs has been almost perfected since its large-scale introduction by Henry Ford in a multitude of industries (see **Volume 1**), allowing for a large array of strategies, processes, and technologies to be implemented. This is considered advantageous for the restructuring of the construction industry according to an OEM model with on-site factories (**Volume 4**) as the final integrator of high-level building components.

Table 1.1. *Analysis framework*

Field of analysis	Analysed factors
Current situation and market shares	Industry shares Manufacturing volumes Raw material consumption
History	Timeline Beginning of industrialized manufacturing Key persons and periods
Range of products	Classification based on geometry Classification based on complexity Classification based on function
Manufacturing methods	Workshop and group-like manufacturing methods Flow–line like, chain-like, and production line–based manufacturing General strategies
Factory layouts	Comparison of various organizational settings and layouts Modularity
Subsystems, end-effectors	Subsystems (e.g., assembly lines, logistics systems, crane systems, handling devices, warehouse systems) Subsystems (e.g., welding, bolting, material gripping, material distribution, material orientation, measuring)
Possibility for industrial customization/personalization	Possibility to customize product by modular approaches Possibility to customize products by automation and robot technology
Emerging topics in the field	Innovations in the field resulting from new manufacturing methods, technologies, and materials
End-of-life strategies	Reverse logistics Remanufacturing Recycling

1.3 Organization of this Volume

The rest of this volume is organized as follows. Chapter 2 focuses on the manufacturing of lower-level components, typically composed of ceramic/brickwork, concrete, wood, steel, glass, and polymers as basic ingredients (building component manufacturing, BCM). Chapter 3 provides examples of the manufacturing of mid-level components (building module manufacturing; e.g., manufacturing of building modules, prefabricated bath modules, or assistance modules, also referred to as building subsystems). Chapters 4 and 5 (large-scale prefabrication, LSP) deal with the off-site manufacturing of complete buildings composed of low-level components, mid-level components, and very high-level components (units). In particular, they focus on systems and kits that are produced (using automation and robot technology) in larger quantities (large-scale).

It must be said that the Japanese LSP industry is far beyond that of other countries in terms of quantities produced, manufacturing technology, and organization, and for this reason, it is described and analysed in detail in a separate chapter (Chapter 5). Japan has the most successful housing prefabrication industry in the world, and has maintained this position for about 40 years. Today, the Japanese

LSP industry manufactures approximately 150,000 entirely prefabricated housing units per annum, with a continuously increasing degree of quality and embedded, advanced technologies. A peak maximum production was reached in 1994, with approximately 600,000 prefabricated housing units. Apart from the large and consistent market share, it is remarkable that the industry supplies higher market segments (rather than the typical lower market segments) with fully customized, earthquake-resistant, high-tech buildings. To be able to provide outstanding quality, almost all manufacturers use automated machines and robot systems in their factories and organize their means of production along a production chain or even production line. The average salaries paid by Japanese prefabrication companies are among the highest in the Japanese general industry. Most Japanese prefabrication companies have no strong roots in the construction industry but rather originate from multinational chemical, electronics, or automotive companies. Currently, the Japanese LSP industry advances directed towards adding and emphasizing complex additional functions and services playing a major role in the country's disaster prevention and disaster management strategy, and developing and delivering (in the role of a kind of super-OEM) entire "smart" cities that are sustainable, affordable, and assistive. Japan's prefabrication industry currently changes the notion of buildings recognized as simple "construction" products towards the notion of buildings recognized as complex high-tech products with completely new, service-oriented value creation potentials – and its advanced manufacturing capability is the backbone for this evolution.

In sequentially proceeding from manufacturing of basic elements, parts and lower-level components to mid- and high-level components, the order of chapters in this volume strictly follows the organizational structure considered as optimal for the deployment of automated/robotic construction, reflected by the OEM model (Section 1.1) and outlined in depth throughout **Volume 1**.

2 Automation and Robotics in Building Component Manufacturing

Building component manufacturing (BCM) is to be distinguished from the manufacturing of medium/high-level building blocks, such as building modules (Chapter 3) and large-scale prefabrication (LSP, see Chapters 4 and 5). This chapter deals with the BCM of parts, assemblies and lower level components and outlines machine systems and manufacturing processes with a particular focus on automation and robot technology. The components and manufacturing systems outlined are based on the processing one of four main materials: ceramic, concrete, wood, and steel. Each is discussed in a separate section. The sections are structured to acquaint readers with the entire manufacturing process, from the processing of raw materials to the production of parts and, finally, to the assembly of those parts into components. The necessary manufacturing methods are outlined to provide an understanding of the basics, identify the specific properties, and describe the variety of processing methods for each material. The chapter focuses on the specific automated machinery, robot systems, end-effectors, jigs, fixtures, workflows, and process layouts necessary within each material category to make, handle, assemble, and process elements and parts into components.

Roughly similar manufacturing methods are used within all of the four material categories in the production and assembly of building components. First, the materials (with the exception of wood) are obtained from a mixture of various basic substances. To achieve permanent union of the mixture of these substances, a transformation process (e.g., curing or hardening process) is required and specific thermal, atmospheric or pressure conditions must be created. Next, thanks to processes such as, for example, extrusion and moulding, the mixed material can be brought into the required form (e.g., basic, raw parts such as bricks, steel bars, and profiles). Since the described transformation of raw materials into parts doesn't take into account the specifications of the future use, further treatment, manipulation and assembly of these simple elements with other elements is required to produce building components with specifications, features, performances and connectors that allow its use in a larger, more complex, modular systems of components, modules, units and buildings.

The process of manufacturing a specific component is beyond raw material transformation completed in many different steps. Briefly stated, these could include cutting, machining, bending, coating, and assembling of elements. Each building component may contain several single elements to be able to offer the desired functions. In BCM (in contrast to conventional construction, where in many cases components are built on the construction site; see **Volume 1, Section 6.5.5**) the manufacturing processes take place off-site, in a structured environment (SE) of a factory, and require a specific manufacturing layout for each material. In this SE advanced manufacturing methods can be used to produce and if necessary customize or personalize the components. In SEs, emerging techniques, such as flexible automation, modular manufacturing systems, flexible jigs/fixtures/end-effectors, robotic systems and the various forms of 3D printing, are ameliorating the industrial production of building components and provide completely new ways of customizing/personalizing components either for subsequent assembly processes (e.g., assembly in LSP [Chapters 4 and 5], processing by STCRs [**Volume 3**] or processing by automated/robotic on-site factories [**Volume 4**]) or the use phase of the building and the customer. Each of the following sections outlines the cutting edge manufacturing technology in each of the four BCM component fields (brickwork/ceramic, concrete, wood, and steel) and makes the reader familiar with manufacturing processes, layouts, and equipment.

2.1 Brickwork- and Ceramics-Based Components

In this section, the concept of a brick refers to the volumetric unit of construction that comprises brickwork building components such as walls and slabs. It also refers to any construction techniques that may emerge from this, not necessarily to the classic clay-based brick. The material could be any kind that is used in the brick element production phase: clay, adobe system, concrete, lime, sand, aggregate, and so on. Advanced machines, automated processes, and robot systems play a large role in the production of low-level parts and components as well as in the assembly of parts and low-level components into higher-level components (e.g., complete wall or ceiling components). Brickwork structures can be preassembled using SEs, advanced machines, automated processes, and robot systems, either off-site or on-site. It is therefore shown in this section that machine systems, logistical aspects, and designs of parts and components must be adapted to the specific use case. An overview of brick and ceramic component manufacturing is also provided. First, traditional brick element fabrication and manipulation techniques are introduced. Next, modern automated or robotic techniques for brick working are described. It should be said that brick and ceramics are one of the oldest construction materials, and some of the original technology may have been diluted in the current field of construction. The latest robotic extrusion and moulding systems have opened up new methods for creating new industrially customized/personalized products from these historic materials. In extensively rewritten and expanded form, Bock and Linner (2009d) builds the basis for Section 2.1.

2.1.1 History and Techniques of Brick and Ceramic Parts Production

Throughout history, brick and ceramic elements have been successfully used as building materials around the world. What are the main factors contributing to this success? Bricks and ceramics are composed primarily of inexpensive and abundant materials such as clay, sand, and lime. Ceramic is defined as a crystallized nonmetallic material (Carter et al., 2013). The crystallization is done through a heating process. Today, a large majority of brick elements are ceramic, although the primitive dried brick, adobe, or mud brick was left to dry without creating a ceramic composition (Schmandt-Besserat, 1977; Stordeur et al., 2007). These bricks don't have an inner crystalline structure, and so they are not considered ceramics. In the very beginning, components used to be manually mixed and moulded with some water. When drying, normally outdoors and under the sun, there was no chemical reaction or liaison between the sand and lime. Once the adobe had been placed in a wall, if it was in contact with a large quantity of water, dried brick elements could become diluted. To improve the inner structural composition, ceramic and fired brick production was introduced (Khan et al., 2013). The elements were fired in kilns to achieve a crystalline nonmetallic substructure that made the element more stable. The more it was vitrified, the easier the material was shaped in fine forms. In that sense, porcelain is considered a highly vitrified type of ceramic (Carbajal et al., 2007). The brick manufacturing technique was further improved by the tunnel kiln (hereafter referred to as a kiln brick) in which the bricks travel through a type of linear oven (Ritchie, 1980). Another step towards the industrialization of brick manufacturing was the development of extruded bricks (Händle, 2007). A clay, sand, and lime mixture is pushed through a die and the profile created is cut to the required measurement. This way, the intended form is shaped much faster than using moulds. Special materials such as quartz, crushed flint, and siliceous sand are added to the initial ingredients. These materials are shaped with water and, to combine the sand and lime chemically, the bricks are subjected to forces in highly pressurized chambers, also called autoclaves. Calcium silicate bricks are created in this way (Bowley, 1994). Fly ash bricks are produced in a similar way, but ash obtained from coal combustion is added to the mixture (Cicek et al., 2007). One of the latest developments in ceramics is ceramic matrix composites (CMCs) or ceramic fiber reinforced ceramic (CFRC) (Chawla, 1998). These materials have excellent heat resistance and toughness, which allows them to be used as parts of brakes on aircraft and sports cars.

2.1.2 Keys and Figures

Brickwork construction has long been a tradition and is still one of the main building techniques used today. For residential buildings, the system is often the preferred first option, as opposed to other options such as concrete or steel. In the construction of residential buildings the use of brickwork is more popular than ever before in Germany thanks to the emergence of mechanized and partly automated assistance equipment (Figure 2.1).

2.1.3 Classification of Ceramic Construction Elements and Brickwork Products

The use of different moulds, dies, and turning tables enables the production of different shapes of ceramic elements. Both simple and customized elements can be assembled to create building components. Below, a classification of the morphology of brickwork and ceramic construction elements is presented.

Figure 2.1. Assistance technology by Layher. (Photo: Layher Bautechnik GmbH)

1. *Simple elements*. All brick unit types are included in this classification. The bricks are available with/without holes, in all dimensional standards and quality for all types of brickwork.
2. *Customized elements*. Analysing another aspect at the volumetric level, customized elements are similar to single elements but with a higher complexity; they are more developed and designed for special functions in the building assembly system.
3. *Building components*. Bricks can be prearranged directly from factory production to build prefabricated building components such as walls and slabs. A particular aspect of the brickwork off-site production is that usually, in contrast with prefabricated concrete production, off-site component production brickwork is not associated directly with the brick production facilities. Usually, the brick production factories are separate and work as suppliers for assemblers.

2.1.4 Manufacturing Methods

We can differentiate between two main phases in the manufacturing process of brickwork/ceramic components. First, the basic element, that is, the brick or ceramic element, is produced from the raw materials. Then those basic elements are joined together to create a more complex (customized) element (which can then be referred to as a component). Ceramic basic elements can be produced in many ways and the production primarily consists of a few basic steps. First, the basic ingredients of the required material are mixed with water in order to create a mixture. Second, this mixture is transformed and brought into shape through a first mechanical manipulation process (e.g., moulded or extruded). All materials are able to take diverse sizes and shapes. Most of the elements for construction are formed either by using moulds or extruded and pressed in open dies. Next, a drying process is completed for eliminating excess water in the element and then the elements are fired. The purpose of firing is to remove the remaining quantity of water and to vitrify the components all together. Today, one of the most advanced shaping techniques is extrusion (Händle, 2007). For producing more advanced basic elements made, for example, from ceramic matrix composites (CMCs) or carbon fibre reinforced ceramics; techniques as, for example, electrophoresis, pyrolysis, sintering, and chemical vapour deposition techniques are used (see, e.g., Krenkel and Berndt, 2005).

From the above-described transformation processes (mixing, shaping, drying, firing, etc.) the resulting basic elements (bricks or ceramic basic elements or building blocks) alone are not construction components. They must be arranged in a specific way and some specific rules and guidelines must be followed for assembly. Many construction forms or components have been developed for the use of brick as a base material, including walls, arches, and domes. The successful combination of these components leads to the construction of entire complex systems/buildings.

1. *Classification regarding the manufacturing place.* For centuries, brickwork has been an on-site, hand-made construction system. Bricks were produced off-site and then transported to the site. A bricklayer controlled the construction procedure by arranging the bricks in their proper place. Sometimes, specific tools were necessary for creating vaults and domes. In recent decades, it has been possible to complete brickwork in specific automated factories, and the on-site-task is therefore to assemble those different components. *Off-site production* refers to the automated/semi-automated production lines of the factories. The bricks come from suppliers and, after sorting and additional quality checks, are introduced into the production line to fabricate building components. Individually planned brickwork panels can be produced in manufacturing plants through semi-automatic production lines, fully automatic production lines, or fully automatic robots under industrialized and highly controlled factory conditions. In the case of on-site production, suppliers deliver the brick elements directly on-site, where the constructors assemble them into the brickwork by using assistive devices and special machineries. The main issue in the case of on-site brick assembly is the time requirement. The brick element is considerably small as a single element compared with precast elements such as brickwork or concrete components. Therefore, an automated or semi-automated brickwork assembly, rather than the classic manual one, would be preferred. Furthermore, using an assistive tool also drastically increases the force used, and so larger dimensions of brick elements can be considered, thus further reducing the work time.

2. *Classification regarding the manufacturing type of the components.* Automated brickwork component manufacturing can be separated into two main groups: horizontal and vertical manufacturing.

    Horizontal brickwork component manufacturing. The Winkelmann robot plant for brickwork is a characteristic example of horizontal brickwork panel production. All single devices are equipped with microsystems, interconnected and part of a systematic logistic network. Computer-aided design (CAD)/computer-aided manufacturing (CAM) ensure efficient data processing between the project and the final product. The basic process of positioning the bricks in a given order is done horizontally on tables or pallets. After delivery, standard bricks on standard pallets are brought into the factory's required processing order by an automated palletizing system with a robot for distribution and arrangement of the bricks. Next, the bricks are taken up by automated depalletizing systems that supply the horizontal brickwork layering robot station. Insertion of reinforcement, infrastructure (plumbing elements, cables, etc.) and plastering is also a stationary process in the factory. Finally, a dried out and finished brickwork panel is delivered to the construction site.

Vertical brickwork component manufacturing. Leonhard Weiss, a robotic brickwork plant, is one characteristic example of vertical brickwork panel production. Many processes, for example, the palletizing and depalletizing processes needed to bring the bricks into factory, are similar to horizontal brickwork production. Yet, the basic process of positioning the bricks in a given order is done vertically layer by layer. High-accuracy robots combined with linear axis systems secure accurate positioning. The automated vertical brickwork layering has particular advantages in terms of the efficient use of factory area. Moreover, the composition and quality of the wall could be improved through vertical processing. The horizontal process has advantages in terms of exact positioning of reinforcement.

2.1.5 Possibilities for Industrial Customization

Rötzer-Ziegel-Element-Werk GmbH is a German company specializing in prefabricated brickwork housing components (Figure 2.2). It is remarkable how the company has developed a system for producing a customized product. In the production line–based and partly automated brickwork factory all of the required building components for a complete house are produced according to customer demands: wall and ceiling components with different levels of complexity that also include considerations for appliances such as electricity, ducts, and shafts including pipes and all connectors. The company uses the horizontal brickwork component manufacturing process.

2.1.6 End-Effectors and Automated Processes

Both ceramic elements and brickwork construction elements can be manufactured and manipulated with complex end-effectors and automated processes. Robotic extruders and handling and positioning devices allow for the automation of brickwork manufacturing.

On-Site Brickwork Manufacturing – Brickwork Handling and Simple Manipulators

Traditional on-site brickwork production by manual means is a difficult task. It requires a large number of personnel, from bricklayers to brick movers.

1. The *Steinherr Assistance System* is a good example of a mechanized, on-site assistance device that supports ergonomically sound working positions, the speed of work, and accuracy or quality with a very simple and flexible solution.

a) Production of brickwork walls on the production line

b) Storage and curing chamber

c) Transport of finished elements by an overhead manipulator

Figure 2.2. Prefabrication and mass customization of brickwork components in the factory. (Photos: Rötzer-Ziegel-Element-Werk GmbH)

The machine also has additional sub-devices for other tasks such as stone saws, measuring instruments, stone shifting grip arms, equipment to apply mortar, and transport carts for materials to facilitate the work and increase production.

2. *Lahyer balancer and handling device for masonry construction* work as flexible devices for brick manipulation that can be distributed over the whole construction site as they are easily assembled, transported, and serviced by construction operators (Figure 2.3). The articulated, multi-joint arm supports every working position and also exculpates often-busy larger cranes from minor tasks. Furthermore, the system is self-balancing and allows free-moving stone positioning without any physical effort. The multi-joint arm has an overhang of 4.5 metres

a) Placing of concrete wall blocks – example 1   b) Placing of concrete
                                                        wall blocks – example 2

Figure 2.3. On-site end-effectors by Layher. (Photos: Layher Bautechnik GmbH)

    for loads up to 400 kilograms. The vertical working position is also variable up to 5 metres and can be wirelessly adjusted by an electronic lifting unit using a remote control device. The balancers can be equipped with various mechanical grippers depending on the type of brickwork or stone used.

3. *Cranes with different end-effectors* lift the various loads and are also movable themselves. This simplifies planning and reduces the costs of a construction project. Machine-driven devices also help to accelerate the workflow. Boecker has developed mini-cranes that can handle loads up to 500 kilograms with a radius of up to 5 metres and a maximum hook height of 6 metres. These systems can function as assistive tools for builders, and their end-effectors are designed for different types of brick units, which permits the handling of almost any kind of brick shape. Another advantage is the size and weight of the mini-cranes; they can be easily transported to the designated places.

4. *Mortar providing elements.* A company called Dunnbett–Mortelschlitten is developing thin bed mortar systems for on-site construction. They are made of stainless steel and ensure an even distribution of mortar on the stones. Using a unilateral side guide, the element runs quickly and accurately and is distinguished by a special grip for an overhead application. Mortar-providing elements are available in all standard wall thicknesses.

5. *Masonry robots for customized construction on-site.* Two robot systems for erecting brickwork on the construction site (on-site) have been developed in the past few decades: the solid material assembly system (SMAS) from Japan and a highly mobile and more advanced bricklaying system later developed at the University of Stuttgart, the robot construction system for computer integrated construction (ROCCO). Both of these approaches have the following characteristics in common:
   - Mobility of the robot system
   - Sensor system for determination of the robot's positions and its environment
   - Offline generation of the robot's motion

- Automatic grappling of the stones from the pallets
- Automatic application of mortar
- Automatic positioning of the bricks

The ROCCO (Figures 2.4 and 2.5; Andres et al., 1994; see also **Volume 1, Chapter 6** for more detailed information on robot-oriented design [ROD] aspects related to this system and in particular its end-effectors) system is mobile and can carry out the brick-layering process independently. The robot removes the bricks from pallets and assumes the proper alignment of the stones using a sensor system. It brings in the mortar automatically. It is possible to program the machine via CAD, similar to the machinery for the production of precast elements. The system determines all necessary intermediate steps independently and the operator must only provide the required materials to the construction site. ROCCO's main purpose was to be a robot system for construction site operations. The CAD representation of the building is used to analyse each wall component and decompose it into the necessary blocks. After this, the optimal positions for the robot are automatically calculated. The next step is the coordination with the pallet's position relative to the robot's position and also with the future positions of the blocks. The final step is to synchronize the robot motion automatically with all previous analysis. One important advantage of this system is that the graphic user interface used on the construction site is intuitive, adjustable, and interactive and requires no special robot programming language to be learnt. The mobile robot begins to execute the application to erect the desired wall. There are two mobile platforms: one, an autonomous vehicle with manipulator without sensors that reaches up to 4 m and the other, a simplified manually operated vehicle able to reach up to 8 m that is equipped with a manipulator that has many sophisticated sensors. In terms of capacity, both are able to lift a payload of 300 kg with a cycle time of 100 seconds. If brickwork prefabrication is done with a great level of accuracy in an off-site factory, it may prove to be more efficient than the ROCCO system, which may incur inaccuracy and positioning faults using a robot on an uncontrolled site. To address this problem, a positioning tool was developed to compensate these inaccuracies using sensors, actuators, and other complementary elements.

The main tool for positioning consists of the gripper itself and a fine positioning device. During the development of the gripper, one requirement was that it had to be able to lift both standard and modified (e.g., sand-lime, clay bricks, or cellular concrete) blocks from the pallet. Finally, different quality control mechanisms were integrated to detect damaged blocks and inaccuracies. With the system outlined, the following work sequences on the construction site evolved. The mobile robot system and block pallets are placed on the floor by the crane in the rough initial positions determined by the working path analysis. After referencing and measuring the position of the vehicle, the robot moves into its working position. After the vehicle is positioned and supported, the actual position will be measured and compared to the planned position. The calculated difference will be used as a compensating value inside the robot control system, which compensates offline programmed commands. The pallets positioned with the construction site crane (quite inaccurately) will be fine-positioned manually in the first step. After the respective block is gripped from the pallet, it will be placed by the manipulator in a rough position on the wall. The

a) Gripper type 1, prototype

b) Mobile platform, prototype

c) Mobile platform, detail

Figure 2.4. Robot construction system for computer-integrated construction (ROCCO).

a) Gripper type 2, front view

b) Gripper type 2, side view

c) Gripper type 2, top view

Figure 2.5. ROCCO's end-effector for positioning concrete elements.

architectural design affects the automated execution of the construction process on-site. Computer-integrated manufacturing (CIM) makes it possible to process all collected data automatically without compromising data consistency.

SMAS (solid material assembly system). Already in the 1980s, Japan had developed a system for automated brick layering of standardized prefabricated elements. With the SMAS (see also **Volume 1, Chapter 6** for more detailed information on ROD aspects related to this system) it is possible to lay concrete-steel stones using a robot. The building components (30 cm × 30 cm × 18 cm in size) are connected by steel joints. A multi-articulated, mobile industrial robot positions the components. Because of the small size of the mobile production unit, it would also be possible to use it on the construction site. Although the system cannot build with normal brick, the structure can serve as a model for a fully automated brick-layering robot that can also be used on a construction site. Components are positioned one by one automatically by the robot. Following the positioning of each component, steel bars are connected to those of adjacent components, also by the robot. The joint type of steel bar for the vertical direction is mechanical and that for the lateral direction is overlapping. Concrete is subsequently grouted from the top of the wall that is erected one story in height (approx. 3 metres). The complementary operating hand is fixed to the mother robot (6 articulation type robot), which has been designed for a wide variety of applications in factory use and a series of experiments for wall erection. SMAS building components have been designed as robot-oriented small sized components to avoid the difficulties that arise when large components must be manipulated on-site. Although masonry structure typically had not been considered as a major structural system in Japan because of the danger of earthquakes, SMAS, when properly reinforced, became recognized as a flexible structural system applicable to customized building designs. A basic component is fitted with steel bars for the reinforcement to be assembled within the stacked wall. For verification of adequacy of the proposed overlap type joint in lateral direction, wall specimens were specially designed and a horizontal loading test was carried out. The result showed that the strength of the proposed joint was comparable to that of an ordinary joint, when properly fabricated.

Off-site Brickwork Manufacturing

There are many mechanized brickwork plants with a low degree of controlled and interconnected processes, and thus they implement a low overall degree of automation. Tools and devices for mechanized brickwork plants include stationary brick-layer workstations with lowering platforms, stationary bricklayer workstations with lifting platforms, mobile floors with lifting platforms, mobile pallets with lifting and lowering platforms, and various auxiliary equipment able to shift the stones.

1. The *Rimaten* is a highly flexible system for manual mechanized brick layering (Figure 2.6). Its hand-guided manipulators support flexible production and moreover are designed for interconnection with automated pallets and other logistic systems. Similar to the Layher Balancer, the Rimaten handling device can be equipped with various grippers and every position on the platform can be reached to position stones with minimal physical effort and high accuracy. In addition, the height of the platform itself can be adjusted to various use cases.

Figure 2.6. Highly flexible stationary lifting platform. (Photo: Rimatem GmbH)

2. The *"Multistone" (Anliker GmbH)* automatic masonry machine consists of a rectangular steel frame with a rotating turning table. The blocks are placed onto a table, which automatically moves them to the correct position and height. Integrated information and communication technology (ICT) ensures that the mortar is applied properly and all subsequent operations are carried out with high accuracy. The "Multistone" can be operated by two people. One person is responsible for controlling the machine and the other one for various preparations, finishing work, and organizational tasks. The blocks arrive stacked on pallets and are placed onto the turning table by the machine operator using a gripper arm. When the blocks have reached their correct position, they are set into the mortar bed that was applied beforehand by the machine. The cutouts for the windows and doors are done before. After the desired wall has been completed, the steel frame is moved on to build the next wall. One or two days later, when the drying process has finished, the electrical wiring is installed. Following this, the walls are plastered and the sills and windows are fitted. Using a low-bed truck, the walls are carried to the construction site and a mobile crane installs them where they are required. With the "Multistone" automatic masonry machine in a simple factory environment, walls can be built up to ten times faster than on the construction site.

Others

Pritschow et al. (1994) have proposed an approach for developing a brick-layering robot. The tasks of the on-site brick-layering robot include depalletizing the bricks or blocks, the application of liant-bond material, and finally the erection of a brickwork wall at a high level of accuracy and quality. The design of the robot was strongly influenced by aspects such as the types of bricks or blocks and the mortar technology used. Furthermore, the same authors have proposed a list of the main functions of the robot:

- Blocks and brick handling with the possibility to adjust according to their sizes
- The ability to detect the material tolerances and adjust them properly
- Using the tool centre point (TCP) system for the calibration of the brick or the respective element position
- Equipped with an automated dispensing instrument for the bonding material
- Designed to deliver cost-effective solutions and adapt to on-site demands and conditions

To meet these requirements, a number of solutions have been presented during the last few years, but each of them addresses only a portion of the requirements.

2.1.7 Factory Layouts

Efficient brickwork manufacturing is based on an optimal factory layout. From brick extrusion to mortar pouring, the process must be completely coordinated. In this sense, it is interesting how Lissmac has developed a completely automated factory for brickwork components. Manufacturing using robots is a process composed of four phases/positions: (1) the lining up of the brick elements, (2) the application of mortar, (3) pallet circulation, and (4) post-processing.

Automated Brickwork Plants

The adoption of CAD/CAM systems, combined with the implementation of advantaged enterprise resource planning (ERP) solutions, has made the implementation of highly and even fully automated brickwork plants possible (Figure 2.7). Highly automated brickwork plants can be characterized by a high degree of interconnected processes and logistics, CAD/CAM system integration, ERP solutions, supply chain management, and subsequent automated processes in the production line.

1. *Brickwork robot plant SÜBA – prototype.* A prototype of an automatic brickwork plant was deployed by SÜBA Bauen und Wohnen Rhein-Neckar GmbH during the 1990s. The production of brickwork panes is appropriate for capacities with 300 square metre net area of brickwork panels – without window and door recesses – within a span of eight hours. The use of CAD in the architect's office made it possible to transfer the large data set needed for the production of brickwork panes directly without manual input over CAM.
2. *Wall partitioning tools* divide architectural walls into the necessary blocks under for the consideration of windows, doors, lintels, and so on. The outputs are the dimensions and positions of each block in the respective wall. During the segmentation procedure, optimization criteria must be considered under hard boundary conditions. The number of nonstandard blocks should be minimized to keep costs low. The dimensions should also be well balanced to keep the waste produced from cutting low.

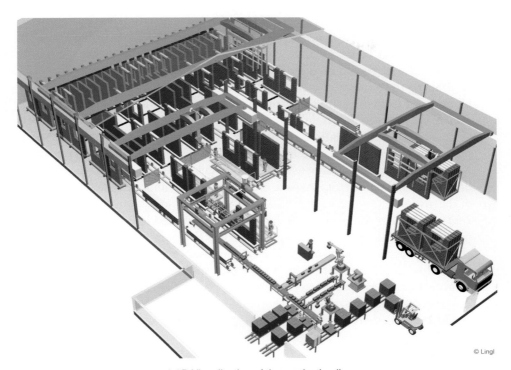

a) 3D Visualization of the production line

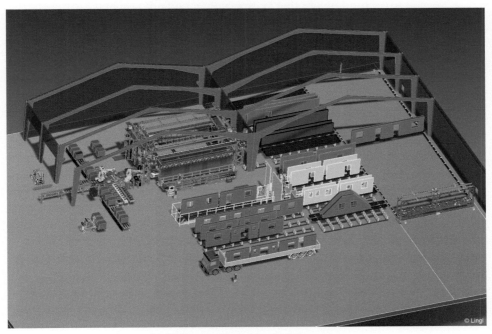

b) 3D Visualization of the logistics system

Figure 2.7. Automated factory for brickwork components. (Photos: Lingl)

3. *Sequence and task planning tools* enable different calculation and optimization procedures. In the first step, the software must determine the possible assembly sequence of the blocks, that is, to generate an assembly sequence graph. In the second step, the optimal sequence concerning the optimization criterion of minimizing the number of vehicle movements while maximizing the number of blocks built from one working position must be determined.

4. *Palletizing* helps to determine the position of the blocks on the pallets as well as the sequence and positions of the pallets on the construction site. Necessary pieces of information include the assembly sequences related to the perspective working points, the dimensions of the blocks and pallets, the position and dimension of the free storage space around the robot, and the specific properties of the gripper.

5. *End-effectors.* A wide variety of end-effectors are used in completing complex operations in the manufacturing processes of brickwork components. They can generally be categorized into end-effectors used in logistics operations such as manipulation, transportation, and storage and end-effectors used in direct manufacturing processes such as palletizing/depalletizing.

6. *Logistics tools* ensure transport and manipulation but also storage of the final products before delivery to the construction site. Some examples of these include
   - Gantry crane for transportation
   - Storage area and racks – a special section in the factory for finishing until the product is dried and ready for transportation
   - Special truck for delivery that is especially modified/designed for transportation of the brickwork components
   - Truck brickwork holder – a special truck subcomponent designed as a brickwork holder for the brickwork products
   - Lime/mortar pouring tools

2.1.8 Emerging Techniques in the Field

Aside from existing automated technologies, new methods for ceramic production and brickwork manufacturing have been developed in the last years. A few examples of these developments are presented in the text that follows. Although it cannot be considered as a brick type or component, the coordination of CAD and robotic fabrication for ceramic products has led to interesting procedures for developing customized products. This was the case in the "Generation of a prototypical high performance ceramic shading system" (Bechthold et al., 2011; Figure 2.8). Thanks to an integrated design workflow using Rhinoceros, each "geometrically complex element" of the shading system can be defined accurately. Following this, redeveloped Grasshopper and Rhinoceros plugins facilitate the configuration and fabrication of moulds. The moulds are fabricated by an ABB robotic arm and pins are used to create the geometry of the mould. Those pins are then covered by an interpolative surface that accepts material deposition. The very same robotic arm manipulates a ceramic extruder that covers the mould. After the mould covering by ceramics, the complex element needs to be finished. A dimensional rectification is achieved through robotic milling.

a) Design of the geometrically complex
ceramic shading system

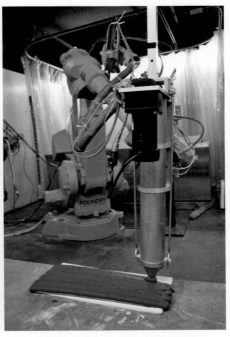

b) Robotically actuated variable mould
configuration using a pin system

c) Robotic ceramic deposition over the
previously configured mould

Figure 2.8. Robotic fabrication workflow for ceramic shading systems. (Photos: Harvard GSD
Design Robotics Group, Prof. Martin Bechthold)

Figure 2.9. The programmed wall, developed at the ETH Zurich. (Photo: Gramazio & Kohler, Architecture and Digital Fabrication, ETH Zurich)

Another interesting example is the "programmed wall" (Figure 2.9). An experimental use of robotics for brickwork has been developed at ETH Zurich by the laboratory of Gramazio and Kohler. The special geometries of the wall were designed previously using CAD software. The very first attempts were without mortar (Bonwetsch et al., 2006).

2.1.9 End-of-Life Strategies

Brick products can be recycled in one of two main ways. The first method is to crush the material and convert it into a powder. This powder can be used as an aggregate in different kinds of mortars or as gravel in exterior grounds or terraces. This process can be easily automated. The second method is to maintain the brick's original form and dimensional properties. After the brickwork is disassembled, the brick element needs to be cleaned to remove the remaining mortar.

Using bricks without mortar could facilitate the ease of reutilization. For this reason, steel reinforced bricks were developed by Dr. Kentaro Yamaguchi, Kyushu University (Khamidi et al., 2004). This new structural system for brickwork construction is based on the distributed unbound pre-stress theory (SRB-DUP). The system is composed of brick elements and a reinforcement grid made of bolts and steel plates (Figure 2.10). The innovation is in the dry assembly of the bricks, that are built with three symmetrical holes, each for the joining system and earthquake reinforcement system composed of steel plates, bolts, and nuts. Aside from its strong

a) Horizontal loading test (WS7 type)   b) Double-story DBHS experimental house

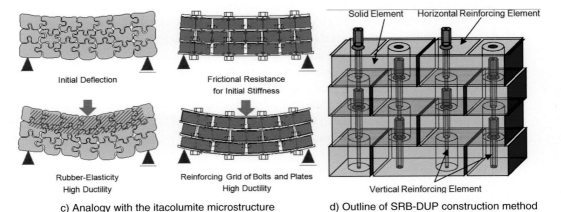

c) Analogy with the itacolumite microstructure   d) Outline of SRB-DUP construction method

Figure 2.10. Steel-reinforced brickwork joined without application of any kind of mortar (Yamaguchi et al., 2005).

behaviour in reacting to earthquake vibrations (an experimental two-story house has already been assembled in Fukuoka), the unbound elements offer the option of disassembly in order to reuse the material. The final appearance of the building resembles that of classic masonry construction. In terms of reusability, this strategy offers new possibilities in that it is possible to disassemble the walls completely and replace them in another location, or simply reuse the brick elements without any losses.

2.2 Concrete-Based Components

Concrete is the most widely used construction material in the world (Nawy, 2008). What is the reason for this? Most likely, the availability of the components and the competitive price of the products play a large role in the popularity of concrete. The use of reinforced concrete components can be considered as industrial. Almost 100% of the fabrication of steel bars, cement, and aggregates is done in industrial plants. The casting procedures, however, are not so industrialized yet. Although reinforced concrete construction is better known as an on-site technique, this section focusses on the prefabrication and automated production of concrete building components. In this section, traditional prefabrication techniques are first explained. Next, the automated manufacturing of concrete components is analysed. A final section describes emerging techniques that hint at the future development of concrete components. In extensively rewritten and expanded form, Bock and Linner (2009b) build the basis for Section 2.2.

2.2.1 History and Techniques of Concrete Prefabrication

Before being produced by manpower, concrete existed as a natural phenomenon, as it has been discovered in Israel as a natural deposit of cement compounds (Schaeffer, 1992). According to Schaeffer, the compounds were formed as the result of a chemical reaction between limestone and oil shale. One of the first known uses of concrete was to construct floors; archaeologists in Syria discovered evidence of its use for this purpose in remains dating to 6500 B.C. In ancient China, concrete was used to keep bamboo together in boats during transport. In ancient times, the Romans used concrete composed of slaked lime mixed with sand for various infrastructure/engineering works such as aqueducts, tunnels, bridges, and so on (Robertson, 1929). In the early 1850s, Jean-Louis Lambot was considered the first creator and user of reinforced concrete for his small boats (Steiger, 1992). He reinforced his boats with iron bars and wire mesh and was known to submit plans and details for using this technology to obtain a patent in France and Belgium in 1956.

Beginning of Prefabrication of Concrete

Precast concrete is a product for the building industry defined as concrete cast in a different location (on-site/off-site) using a predesigned structure. The prefabricated concrete is made utilizing standardization and mass production techniques to achieve maximum quality and efficiency. Modern buildings using precast concrete were first developed in Liverpool around 1905 (Sutherland et al., 2001) by an engineer named John Alexander Brodie. He was one of the pioneers supporting the idea of prefabricated housing and mass production building. He presented his project "Cheap Cottages Exhibition" at Letchworth, promoting precast reinforced concrete slabs as a solution for easy housing construction (Powers, 2007). Grosvenor Atterbury continued Brodie's idea and was one of the first people to come up with the idea of element standardization (Atterbury, 1920). Atterbury designed a house that used approximately 170 precast concrete panels, produced off-site and assembled by crane (Bergdoll et al., 2008). In the assembly operation he split the processes into two steps: (1) formwork to truck and (2) truck to crane. Atterbury's system has become an inspirational resource for modern architects such as Ernst May (Bergdoll et al., 2008). May's main work, *Rundling* in the Römerstadt, Frankfurt, used prefabricated concrete panels for housing (Schleuning, 2000). A keen interest in prefabricated concrete began in the 1950s when the metabolism movement considered it the optimum material for their visionary urban and architectural development. Works such as Nakagin Capsule Tower by Kisho Kurokawa, Mega City Planning for Tokyo by Kenzo Tange (Abley & Schwinge, 2006), and Habitat 67 in Montréal by Moshe Safdie (Murphy, 2009) are proof of the benefits of precast concrete benefits offered by its modularity and flexibility. The use of precast concrete garners special consideration here because of the special geological conditions (earthquake instability) and

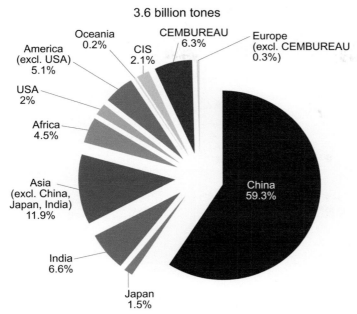

Figure 2.11. World cement production 2012, by region and main countries. (Authors' interpretation based on CEMBUREAU, 2013)

for political reasons (all the countries were under Soviet Union political influence, and therefore a common building standard could be adopted). Countries such as Kazakhstan, Kyrgyzstan, Russian Federation, Uzbekistan, Serbia, and Montenegro have produced precast concrete elements for building in both industrial and civil construction fields using different systems: large-panel, frame, and slab-column systems with walls (Brzev & Guevara-Perez, 2013).

2.2.2 Keys and Figures

Concrete production is primarily linked to its composite structure, the main elements of which are aggregate, cement, and water. The aggregate is usually raw material (natural gravel) but substitutes can also be used, especially in countries where this natural resource is not readily available. Japan, for instance, has exploited to exhaustion its natural gravel resources and is now importing it from Taiwan to produce concrete (Dunne et al., 1980). Cement has a long history originating in the Roman ages and its industry is directly linked with the concrete production industry (cement is produced mostly for creating concrete). It is also linked with the construction field, and therefore cement and gravel statistics are useful tools in estimating concrete production. Prefabricated concrete can be considered a good resource in the efficient production of building elements, a method to obtain products of high quality and performance with a more efficient use of the materials. Also, new innovations in building materials, such as ultra-high-performance concrete (UHPC), are better economical and efficient approaches as new prefabricated building components have smaller sections than the previous ones using standard concrete.

According to the European cement association CEMBUREAU (CEMBUREAU, 2013; see also Figures 2.11 and 2.12), the major cement producers are

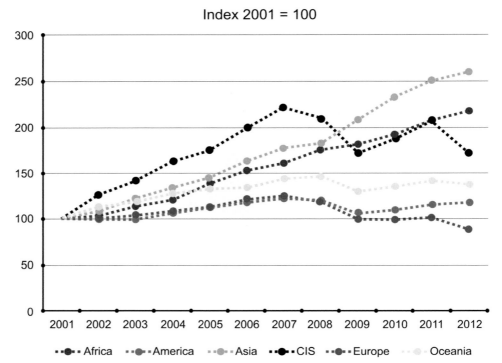

Figure 2.12. World cement production by region – evolution 2001–2012. (Authors' interpretation based on CEMBUREAU, 2013)

generally China and other Asian countries (excluding India and Japan). China is known as a massive producer for both domestic use and export purposes and it is consuming a large amount of natural raw resources for its industry. CEMBUREAU affiliate members also contribute, with 7.3% of the world cement production, and the association states that world cement production is increasing again after a major stagnation between 2008 and 2009. Although European states, including CEMBUREAU members, are facing a decrease of production, with some hopes of recovery, the main producers such as China and other Asian countries are increasing their production. The European Ready Mixed Concrete Organization (ERMCO, 2014) also confirms the decrease of cement production, mainly for the European producers, between 2008 and 2010 although there are some signs of recovery in Germany and Finland. The industry in the United States has remained constant for the last couple of years, as shown later, while countries such as Russia and Japan are reducing their production. Israel and especially Turkey are experiencing considerable production development within the cement industry with increases of 1.6 million m^3 ready-mixed concrete and 15 million m^3 ready-mixed concrete in 2010 respectively.

2.2.3 Classification of Precast Concrete Products

The variety of concrete prefabricated elements and components is large, and so, the next classifications intend to clarify the heterogeneous manufactured components. The simplest components can be considered as linear or 1D. The complexity grows

Table 2.1. *Classification based on the geometry and complexity of concrete products*

COMPONENTS (1D)	PANELIZED CONSTRUCTION (2D)	VOLUMETRIC CONSTRUCTION (3D)
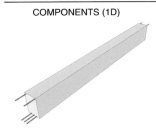	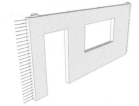	
a) The simplest precast concrete elements, usually one-direction (linear) elements	b) Most suitable for building programs such as hotels, student housing	c) Mostly room modules (individual parts) or panels joined together in factory
Examples: beams, columns, and stairs	Advantage: repetition of similar elements generates mass production – high cost, efficiency	Advantage: high quality of installations such as wiring, complete functional module units such as bathrooms, kitchens

as the dimensional complexity of the component becomes 2D or 3D. With the complexity of the component also the complexity of the manufacturing system rises.

Classification of the Geometry and Complexity of Concrete Products

The building industry has invested a large amount of money during the last years in precast concrete technology, improving both materials and methods. In addition, 3D building information technology (BIM) has helped the construction industry to develop better products allowing more flexibility and degrees of freedom for design. In general, there are three types of categories for the geometry of precast concrete elements: components (1D), panels (2D), and volumetric (3D) elements (see also Table 2.1).

In terms of complexity, precast concrete components are the simplest elements of construction. Products such as precast concrete beams, columns, and stairs have been mass produced for a long time and their benefits, in terms of both economic and structural performance, are proven. On the other hand, panelised construction panels have their own market. Developed for commercial applications such as hotels, student facilities, and special housing, they require building programs. This implies repetitive elements to make mass production possible, and through this to achieve a high effective production cost. In terms of benefits, they provide a fast build time. Although at the same time there is also the necessity of additional treatments such as the over-cladding of panels, which, for some producers could be considered a disadvantage in terms of costs and quality. Volumetric structures are mostly room modules cast as one part or panels joined together in a factory. They have a wide field of use where repetitive modules are needed – from housing to hotels, student housing, or prisons. The most significant benefit of these structures is the high quality of installation, which is of great importance in service–intensive buildings. Entire bathrooms or kitchens can be precisely installed in a factory and delivered within the room module. However, the transportation costs and restrictions of size or weight prevent their wide popularity among building companies.

Table 2.2. *Classification based on the architectural/engineering function of the concrete product*

LARGE PANEL SYSTEMS	FRAME PRECAST SYSTEMS	SLAB COLUMN SYSTEMS
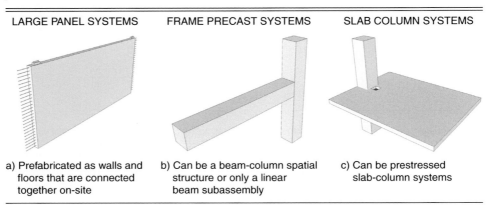		
a) Prefabricated as walls and floors that are connected together on-site	b) Can be a beam-column spatial structure or only a linear beam subassembly	c) Can be prestressed slab-column systems

Classification of the Architectural/Engineering Function of the Concrete Product

Precast structures can be classified into three different categories depending on the load-bearing structure: (1) large panel systems, (2) frame systems, and (3) slab-column systems with walls and mixed systems (see also Table 2.2). Large panel systems are used for multistory buildings and facilities. The panels are prefabricated as walls and floors, which are later connected together on-site, in both horizontal and vertical directions, forming the room units of the buildings. Frame precast systems are systems that can be produced in two ways: a beam-column spatial structure or simply a linear beam subassembly. The advantage of these systems is that the final product is easy to handle and therefore also easy to implement. The advantage of slab column systems with walls and mixed systems is that the walls are designed for shear stress and therefore can carry lateral forces. They are available as pre-stressed slab-column systems or lift-slab with walls systems.

2.2.4 Manufacturing Methods for Precast Concrete

In recent years, concrete prefabrication technology has become more oriented to the industrialized automated methods of production. Automated production methods such as robot-assisted production and control via the latest CAD/CAM technology have been imported from other industries such as automobile, airplane, and shipbuilding.

There are primarily two types of industrialized production methods for prefabricated concrete. The first type is using a stationary single formwork. Such production methods use mobile workflows (cleaning, pouring/casting, and moving the final product to the storage) and the formworks are stationary during the whole process of prefabrication. The second production method uses a mobile formwork. The basic characteristics of this method are opposite those of the stationary method: the work posts are stationary and the formwork is mobile, moving from post to post on the assembly line.

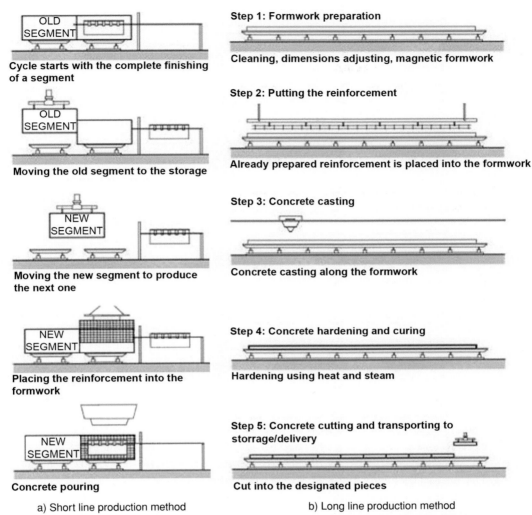

Figure 2.13. Basic functional principle for stationary short/long line production method. (Authors' interpretation based on Girmscheid, 2010)

Stationary formwork production methods: In short line production methods (Figure 2.13(a)), all of the tools, materials, and machines are brought to a specific point following the characteristic sequence of production. One single formwork is used to build the precast concrete element. In contrast with short line production, the stationary long line production method (Figure 2.13(b)) implies that the equipment and materials required for all production steps operate along/above the production line to acquire specific sequences of production.

Mobile formwork production methods – circuit production (Figure 2.14). This production method is used for precast concrete elements such as big slabs, or walls (in general 2D concrete elements). Production involves transporting the element into the factory on mobile tables/platforms or by conveyors from one specific process to another.

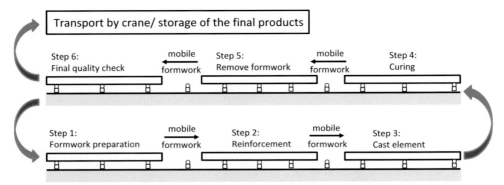

Figure 2.14. Basic working principle for circuit production. (Authors' interpretation based on Girmscheid, 2010)

2.2.5 Possibilities for Industrial Customization

Every customized concrete prefabricated product requires a specifically defined formwork. The recently developed in foam-based formworks are a great advance for creating "tailor-made" concrete products. One example of this is the project *Neuer Zollhof* by Frank O. Gehry in Dusseldorf (see also Bock & Linner, 2009b). The total construction time of this project was three years (1996–1999). The building complex consists of three buildings: house A, house B, and house C. The perimeter walls of house A are planar and vertically inclined prefabricated elements. These elements proved to be the least costly solution for these extraordinary shapes. The geometry of the pieces was designed digitally and then built in three dimensions. The 2D drawings for the formwork plans were automatically generated by CATIA-CAD, a software program designed for airplane construction. The plant produced the necessary pieces according to those drawings. Almost every part is unique. The precast elements had a thickness of 25 centimetres and were approximately 6 metres high, 4 metres wide, and weighed 9 metric tons. All parts were assembled by crane except in a few situations, where a lorry mounted telescopic crane was needed.

The clinker facade required extensive details and good craftsmanship. House B consists of 355 prefabricated non-structural perimeter wall elements supported by cast-in-place concrete slabs. Each floor was split into single elements according to assembly and structural needs and the data were then reconverted into CATIA to be delivered to a milling shop. The freeform incline surface of the cast-in-place walls of house C were built in a similar way as the prefabricated elements of house B. The concrete parts were digitally generated so that the computerized numerical control (CNC)-milled and form-defining polystyrene pieces fit between the planar formworks. On-site, the pieces were integrated into the regular formwork. The walls are all cast-in-place concrete. These specific polystyrene formworks were produced by CNC milling machines. Using the formworks, the 18 centimetre thick precast concrete elements were manufactured and delivered to the construction site "just in time".

Another interesting case is the Bodenkiesel Media House (see also Bock & Linner, 2009b). The organic-shaped space was developed using a highly complex and unique construction method. The unusual building with three-dimensional bending walls is based on a wood-steel structure clad with 124 glass-fibre concrete facade elements with a maximum height of 5.30 metres, maximum width of 4 metres, and thickness of 25 millimetres. The facade elements are thin and flexible but at the same time, stable glass-fibre concrete is fixed to a glued-laminated timber girder combined with a steel frame structure. To manufacture the facade elements, the specialists decided to first build a 1:1 Styrofoam model. Using the digital model as a basis, the heights and sections for actual construction were calculated. The production of the prefabricated glass-fibre concrete elements was a highly complex procedure. The Styrofoam igloo was digitally scanned and then split into 124 pieces. The single pieces simultaneously became formworks for the manufacturing process.

Construction by conventional methods using steel reinforcement would indeed have been impossible; however the 2.5 centimetre thick glass-fibre concrete and the unconventional engineering method made it possible. Every facade element became unique. To ensure the same colour for each element, the manufacturer needed to pay special heed to the sequencing of the mixture, the temperature, and the machined mixing in the production facilities.

2.2.6 Equipment and End-Effectors for Automated Production

A variety of different equipment and end-effectors (Table 2.3; Figure 2.15) facilitate the manufacturing process of prefab concrete. An overview of the most important equipment and end-effectors (exemplarily from the company Weckenmann; Weckenmann, 2015) is given in this subsection.

1. *Pallets, moulding systems.* Pallets (component carriers) are designed for use in the manufacturing of both two-dimensional precast concrete ceiling elements and three-dimensional precast concrete elements.
2. *Cleaning, measuring, and oiling systems/robots.* The setup times for production lines and pallets are decisively reduced by such systems. Cleaning and plotting robots are used for a variety of tasks such as picking up, insertion of latitudinal anchors, cleaning of pallets, full-scale plotting of elements, and installation of latitudinal anchors.
3. *Robotic shuttering systems.* A wide range of slab geometries can be processed by such systems. Fully automatic robot systems were developed for positioning magnets and shutter units and plotting all geometrical slab information as required (e.g., Twin-Z-robot, see also Figure 2.19). Robotic shuttering systems are composed of the following components:
   - (Four-axis) Gantry robot
   - Feeding belt for shutter profiles and magnets
   - Cleaner for shutter profiles and for magnets
   - Identification belt for shutter profiles
   - Congestion roller conveyor for magnets
   - Magazine for shutter profiles
4. *Semi-automated and robotic concrete distributors.* Gantry concrete spreaders for feeding precasting lanes and semi-automated concrete distributors are used to

Table 2.3. *End-effectors for concrete component production*

a) Gantry concrete spreader

b) Automated low-noise vibrating station

c) Battery mould

d) Curing chamber

e) Demoulding, depalletizing device

f) Cleaning and oiling

Photos by Weckenmann Anlagentechnik GmbH & Co. KG.

Figure 2.15. Robotic leveller. (Photo: Weckenmann Anlagentechnik GmbH & Co. KG)

perform different activities. In general, robotic concrete distributors have the following basic elements:

- Equipment: Transverse moving gear, stepless spiral roller drive, distribution roller, local lighting, central lubrication unit, screed, roughening device, lined bucket interior, lane vibrator, hydraulic valve with adjustment.
- Screed bar for smoothing the surface during the concreting of solid walls. The bar has two external vibrators for producing an extremely smooth surface.

5. *Automated low-noise vibrating stations.* The automatic concrete distributor is installed above the vibration station.

6. *Low curing chamber, rack-storage crane.* The low curing chamber (from Weckenmann, for example) is composed of 4 racks with a maximum of up to 50 pallets storage boxes.

7. *Transport and storage system.* Using automated systems (portal vehicles, double-support bridges with longitudinal and horizontal running mechanism, or extra devices for existing cranes), large-scale concrete slabs can be lifted, transported, and stacked.

8. *Formwork end-effectors*: Formwork end-effectors (such as the ones used, e.g., by Twin-Z-Robot from Weckenmann) are designed for all operations required to process the formwork unit:
- Cleaning and varnishing the formwork
- Inspecting, checking, and plotting the positions for the magnetic steel bars
- Picking up and placing the magnetic steel bars for formwork
- Placing insulation forming precast concrete sandwich walls and slabs
- Concrete panel painting

PROCESS AND EQUIPMENTS
1. Pallet cleaning and release agent spraying equipment
2. Plotter, shuttering, and stripping robot
3. Mesh welding plant
4. Reinforcement placing robot
5. Concrete distributer + Compacting equipment
6. Turning equipment
7. Pallet stacker
8. Floating equipment
9. Tilting equipment
10. Run-off carriage

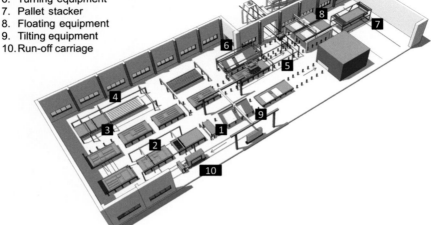

Figure 2.16. Typical indoor concrete factory layout. (Photo: EBAWE Anlagentechnik GmbH)

2.2.7 Factory Production Layouts

Technologies infused into the construction field from other industries such as the automotive industry are the clearest proof of a shifting in building strategy. Fully automated fabrication processes, which are now possible, are implemented by factories to adapt their products to the new higher requirements of the market. Figure 2.16 visualizes a typical indoor factory layout and Figure 2.17 a typical outdoor factory layout.

2.2.8 Emerging Techniques in the Field

Prefabricated concrete component production is constantly being optimized. In particular, in combination with prefabrication by machines, automated systems and robots in SEs, following innovations have potential:

1. *Polymer optical fibre (POF) sensors for high strains measurement.* One research development using optical fibres is the POF sensor system used to measure high strains in reinforced concrete and steel structures subjected to dynamic load conditions, such as those imposed by earthquake loads (Lenke et al., 2009). Generally, POF sensor systems have the potential to offer a larger strain range measurement capability along with improved long-term survivability (Abdi et al., 2008).
2. *Insulation process application robot (IPAR) from Sommer (Figure 2.18).* The German company Sommer has developed an automated set of processes that

PROCESS AND EQUIPMENTS
1. Concrete mixer
2. Mixer to add accelerator for concrete
3. Concrete pump
4. Setting reinforcing bars
5. Concrete casting machine
6. Automated form
7. Reinforced bars assembly system
8. Component stock yard
9. Crane

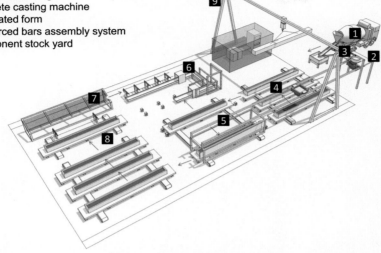

Figure 2.17. Concept for an outdoor prefabricated concrete factory by Takenaka (visualization according to Construction Robot System Catalogue in Japan, 1999).

can perform multiple tasks to obtain insulated prefabricated concrete products: slabs, double walls, and sandwiched wall. The main advantages of this system are that (1) there is no influence on cycle times (linear non-interruptive processes) and (2) the whole automation can be inserted as a production line into existing precast concrete plants. The precast concrete industry finds this new technology encouraging, as it provides a new class of prefabricated products: insulated precast concrete products.

3. *Robotic high-speed shuttering systems.* The German company Weckenmann has been developing formwork robots for precast concrete factories since 1992, and has a portfolio of more than 80 precast concrete plants. Their method involves placing the shuttering steel profiles without error using information directly from the CAD data. Robots can perform different operations such as sorting/placing the profiles and placing different additional components such as magnetic boxes, electrical sockets, inserts, and special face brickwork if required. They can activate the magnets that are inserted into the formwork profiles. Shuttering/deshuttering with already scanned profiles is also possible. New robots such as the Twin-Z-Robot (Figure 2.19) can operate up to two times faster than conventional systems due to new kinematic compositions.

4. *3D concrete printing (Figure 2.20).* Dr. Richard Buswell and Prof. Simon Austin of The School of Civil and Building Engineering at Loughborough University, United Kingdom, have demonstrated the viability of the manufacturing process known as 3D concrete printing (3DCP). This process is described by Lim et al. (2012). 3DCP is a large-scale additive manufacturing process designed to produce full-scale components for construction. Some of the benefits of this method

a) Placement of formwork before pouring concrete

b) Placement of insulation

Figure 2.18. Sommer IPAR system, multitask operations: pouring, drilling, and planning the wall connectors. (Photos : Sommer-Landshut)

c) Removal of formwork after concrete pouring

d) Detail of end-effector

Figure 2.18 (*continued*)

of manufacturing are that it empowers architects and building designers with a large degree of freedom and the possibility of producing a building element directly from a digital model. Printed concrete has 90 to 95% of the strength of the conventionally cast equivalent.

Figure 2.19. Robotic high-speed shuttering system, Twin-Z-Robot. (Photo: Weckenmann Anlagentechnik GmbH & Co. KG)

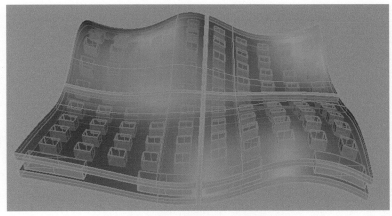

a) 3D modelling of the required concrete model

b) Degree of detailing inside the 3D printed concrete model

c) 3D printed façade element

Figure 2.20. 3D concrete printing (3DCP) technology developed at Loughborough University.

5. *Concrete fulfilling tubular (CFT)* is ultra-high-strength concrete also known for its high quality and resistance to fire (Kawano et al., 2002). Filling a steel tube with concrete provides synergetic strength to the CFT: the steel tube helps bind the concrete and the concrete prevents the steel tube from buckling. The "super strong CFT" is made of extremely strong concrete and steel metal, so the shape is slimmer than that of the standard CFT. This enables the building of huge spaces such as an atrium with far fewer columns.

6. *Lunar pad development (lunar-crete).* The Department of Civil Engineering, Hanyang University, Seoul, Korea has conducted the "Experimental Research for Construction of a Lunar Landing Pad" using a special type of concrete that could be produced without water, using lunar dust and 5% of a special polymer brought from Earth (Lee et al., 2011). A prototype for this kind of concrete production has been built and the whole process has been simulated. The results weren't positive because no solution was found for procuring energy to heat the formwork. Also, the final resulting material itself didn't have sufficient strength and resistance.

7. *Concrete with optical fibres.* Dupont Lightstone (Dupont Lightstone, 2014) has invented a new screen technology based on concrete and fibre optics. It is a ground-breaking type of digital signage that projects an interactive image

d) 3D concrete printing equipment in the civil and building
engineering laboratories at Loughborough University

Figure 2.20 (*continued*)

directly onto surfaces such as floors, sidewalks, or other exposed surfaces such
as facades at street level.

8. *LiTraCon* has embedded optical fibres within poured concrete blocks, creating
(partially) transparent concrete that is just as strong as its non-transparent coun-
terpart (Zhou et al., 2009). By embedding optical fibres into the concrete and
controlling the position of each fibre, the image is projected through to the other
side, where it appears.

9. *Eska glowing fibres and concrete.* Similar to LiTraCon, Eska is a product developed by a company named Luccon (Luccon, 2014). It is based on concrete and its optical fibres. The product was used by the architect Kengo Kuma, who first presented his designs at Tokyo Fiber Senseware in 2009. The material is a compound of plastic optical fibres, glowing fibres called Eska, and concrete. The result is an interesting effect with mixing levels of transparency according to the composition of the concrete. The plastic fibres embedded in concrete can be produced easily and in large diameters and are therefore considered as future support for the communication industry.

2.2.9 End-of-Life Strategies

There are many important reasons to consider concrete recycling as an important process in the ready-mixed and precast concrete industry. The production process is very complex owing to concrete's composite structure and is directly linked to CO_2 emissions and natural resource consumption. Natural gravel has nearly or totally disappeared in some countries such as Japan, Sweden, and according to the European research project Eco Serve Network (ECO-SERVE, 2014), Europe in general. Natural sand and gravel resources are also being depleted; therefore, the new direction must be oriented towards crushed aggregate and concrete recycling.

2.3 Wood-Based Components

Brushes and branches were probably the most primitive construction materials for the first builders. Since then, wood has remained necessary for construction and it is not audacious to state that timber has played some role in the majority of buildings for many centuries. Special guilds have preserved the tradition of woodworking and wood parts/components manufacturing. For quite some time, woodworking generally remained a manual procedure. Recently, industrial techniques for creating engineered wood and special machinery for manipulating timber components have opened new possibilities. Furthermore, the growing concern regarding environmental issues has pushed for a more sustainable and balanced level of timber consumption. This section analyses the latest trends and future directions of wood component manufacturing. In extensively rewritten and expanded form, Bock and Linner (2009c) build the basis for Section 2.3.

2.3.1 History and Techniques of Wood/Timber Construction and Prefabrication

In Japan, the prefabrication of timber elements, first for temples and later for houses, has a strong culture. Advanced fabrication of wood elements has also reached a high level in some European and Scandinavian areas. Wooden prefabrication provides many advantages, as the manufacturing of different components occurs nearly without manpower and automation has already reached a high standard. There are various methods for wooden construction, all differing in the framing structure, as follows.

- The oldest method is wooden log construction, which as a massive construction method, is hardly used in advanced construction today. The potential prefabrication degree is fairly low and is restricted to the addition of the individual beam.
- The second method is timber frame construction, which deals with pure technical carpentered connections and precise joints. This method of construction suits prefabrication in the sense that machine manufactured connecting pieces are a possibility.
- The third method is post construction characterized by several bar-shaped constructional components with the bracing accomplished through horizontally and vertically nailed boards. In this case, prefabrication is also limited to the manufacturing of single rods. A congener to this method is wooden frame construction. This differs from the post-construction method in that it has a larger span width and a primary structure that is completely separate from the secondary structure, thus allowing a freer arrangement of the floor plan. In this case, the level of prefabrication remains the same.
- One of the most widely used methods of construction in Europe is timber panel construction. This method also works with the use of wooden framing; however, in contrast to the aforementioned procedures, the bracing occurs using externally placed wooden plates. This brings about a higher degree of prefabrication.

2.3.2 Keys and Figures

Overall, lumber production and consumption saw a dramatic increase in the 21st century (Figure 2.21). A distinction should be noted between two main product tendencies in the overall wood production – sawn wood production and highly processed/engineered wood production.

Sawn wood production has remained quite stable in the last 50 years. In fact, there was a considerable fall in sawn timber production during the 1990s because of a lack of lumber availability and the improvement of other construction materials such as steel and concrete (UNECE, 2005). Thanks to reforestation policies and the use of "glulam" wood, the falling tendency has stabilized and even begun to increase slightly. On the other hand, processed/engineered timber production has been rising since the 1960s. This kind of wood is created from chopped or sliced elements. The pathologies related to wooden elements have been minimized and the possibilities of machined elements are growing. Therefore, it is not bold to say that, although far from the utilization of the traditional sawn timber, new timber products have an interesting and prospective future.

2.3.3 Classification of Products

In the current market, two main categories of wooden products exist. One is traditional solid wood, which is sawn, dried, and typically protected timber. The other group is processed, engineered wood. This kind of wood has previously been both physically and chemically processed. Each material has different characteristics and purposes. The material is normally prescribed according to various specific

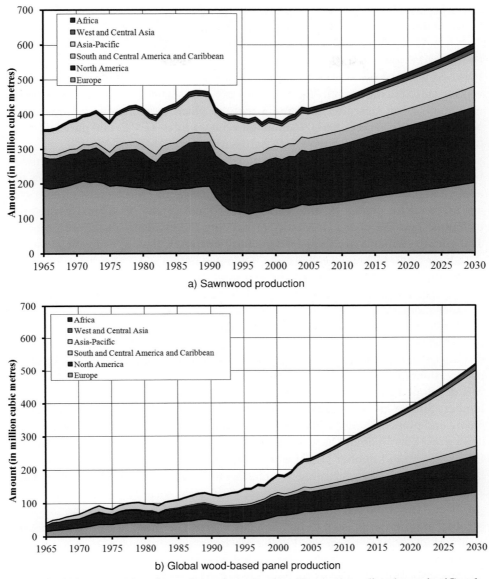

Figure 2.21. Sawn wood and wood panel production. Past and predicted trends. (Graphs provided by Food and Agriculture Organization of the United Nations, according to FAO, 2009)

circumstances. For instance, particleboard is not recommended for outdoor or structural applications because of its engineered composition.

Solid Wood Types

To provide a brief classification, it can be said that softwood comes from gymnosperm species, which are mostly conifers. The growing time is relatively fast and the hardness and strength are normally lower than those of hardwood. These come from angiosperms. Oak and birch are probably the most used of the hardwood species. Tropical timber can also be considered hardwood.

Processed/Engineered Wood

Plywood was first developed by the Egyptians in 2000 B.C. In 1865, the first patent was registered in the United States. The process starts with laminating several layers of lumber. After this, thin sheets of wood are glued by melamine adhesives and pressed at 140°C (Kües, 2007). Laminated veneer timber (LVT) and parallel strand lumber (PSL) are products similar to hardwood.

The production process of glued laminated timber (Chugg, 1964) can be said to be similar to (but simpler than) that of plywood processing. Normally, heat is not applied in the process and the wooden sections are larger. Mechanical or hydraulic jigs are used for gluing. A wide variety of geometries can be achieved using this technique because of the possibility of bending the elements during the process. The glulam elements are used for structural purposes, that is, as linear elements. Although, a new technique called cross-laminated timber (CLT) has been developed for creating structural panels out of sawn wood (Gagnon et al., 2011). Most of the time, finger joints are used for both glulam and CLT to assemble different sawn wood elements.

When harvesting trees for timber, some cannot be manipulated in a saw because of their small size. This lumber waste can be used to create chips and flakes. These elements are the base materials for creating new construction elements. For instance, particle board is composed of wood chips and specific gluing elements. This product was developed in 1940 by Max Himmelheber to address the shortage of natural lumber. Similar to particle board, oriented strand board (OSB) is fabricated by gluing and pressing lumber flakes (Wentworth, 1982). The finishing is rougher than in plywood but obtaining the base material, which is, the strand, requires less timber. Laminated strand lumber (LSL) is a further developed product, where thin OSB layers are glued again, creating a material offering structural properties (Moses et al., 2003). Wood waste is also used for fabricating medium-density fibreboards (MDFs). After wood waste is converted into powder or flour, a mixture of adhesives is rolled in different presses to obtain a flat board. Therefore, it can be considered as a sort of extrusion process. Lately, plastic wood or wood–plastic composites have emerged as a solution for exterior finishes (Rizvi et al., 2002). Thanks to the thermoplastic polymer and wood flour mixture, the performance of this material in exigent environments is better than that of the previous examples. Like MDF, it can be extruded and complex profile sections are possible. However, the structural strength in this case is lower.

2.3.4 Manufacturing Methods

Traditional wood manufacturing is defined by specific customs. From cutting a tree to producing the final wood product, timber must be processed following certain rules. Experience and traditions in this field vary greatly. Guilds, for example, have safeguarded specific woodworking knowledge since the Middle Ages (Wemger, 1984).

Once the tree trunk is lumbered, it must be sawn as soon as possible. After this, the excess inner humidity of the sawn timber should be removed to avoid future dimensional changes. There are two primary methods for drying wood. The traditional or natural way is storing the material in a dry environment (Kahn, 1986).

The modern or artificial method of drying wood is the use of kilns. Once the timber is ready to work with, an approximation of the final measurement must be made. In other words, the sawn wood needs to be cut again and planed (Lye, 1985). The planar dimensions of the wooden element must be adjusted so as to proceed with accuracy. If the required construction element is simple, for example, a beam, the wooden component can be placed in its final location. More complex elements, however, require shaping. This is the case for window frames, flooring elements, and others. Special milling machinery is used for achieving the designed form. Wood is a flexible material. Bending has been used traditionally in ship construction, the furniture industry, and musical instrument craftsmanship (Neison, 2010). Since the mid-19th century, it has been possible to bend wooden profiles thanks to the use of steam chambers.

Basically, all assemblies need to be previously machined and glued, like in finger joints. Especially in glulam structures, but also in furniture, steel joinery offers extra strength for strengthening the union of different elements (Aicher, 2013). Finally, wood needs protection against pathologies caused by insects and fungi, humidity, and other factors.

2.3.5 Possibilities for Industrial Customization

The manufacturing of prefabricated construction elements has, for the most part, been taken over by machines. This addresses the issue of how the data exchange between units can be optimized to achieve the designated goal. This exchange of information also plays a central role in the control of construction robots. Today, this is accomplished through CAD/CAM systems (Tamke, 2009). This means that data are transmitted directly from the drawing program (CAD) of the planner and relayed through CAM programs to the individual machine that fabricates components or provides a robot with its orders. CAM is synonymous with computer-controlled fabrication. These programs are connected directly to CNC machines that obtain the needed information for the next step.

The goal behind this system is the technical informational integration of all aspects concerning a structure. This requires connected, planned automation, optimization of construction sequencing, and the integration of logistics and facility management. A file format that can transfer these types of information is needed to accomplish this complex task. However, the present technology has not yet matured to this level and works primarily within the Drawing Exchange Format (DXF) and Drawing (DWG) file formats.

The problems with these formats lie in the insufficient three-dimensional information that cannot be object oriented, thus complicating the work with CAD/CAM systems. This denotes an additional complexity because the data must be revised and entered once again. For this reason, a new file format named the Industry

Foundation Classes (IFC) is being developed. This new format is capable of three-dimensional data transfer without the additional step of re-entering data. Research in this direction is still ongoing; however, the tendency of computer control invariably persists.

Wooden Framing Method and Multilevel Automated Joining

The mechanical production of wooden components was developed in the 1980s by the Japanese, who created a system that could perform multiple working steps simultaneously. This machine was later referred to in Germany as a joinery machine. The machine is CNC controlled and characterized by its high performance and low-level program complexity.

The most well-known multilevel joinery machine in Germany is the K1, manufactured by Hundegger, and is configured to operate with wooden profile sizes ranging from 20×40 millimetres to 300×450 millimetres. This particular machine operates with the principle of a fully automatic single feeding mechanism to handle both batches and single wooden parts without loss of performance. However, this kind of machine is suitable only for bar-shaped construction components, meaning, in this case, only for timber frame construction, post construction, and timber panel construction. The prefabrication is therefore restricted to the manufacturing of individual components for the carrying structure, thus assigning the procedure a low degree of prefabrication. A relatively time-consuming assembly cannot be avoided. The machine does, however, contain numerous features including sawing, planing, drilling, and moulding. A sliding carriage slides the baulks into the desired positions to complete the correct working steps in the correct sequence.

The spectrum of fully automated processes covers a wide range of possibilities extending from normal rafters and valley rafters with hexagonal cross sections to end-grain and log house processing to moulded and adorned fence posts. The machine is able to complete complex tasks and working steps such as connection manufacturing and straight-lined drilling without any difficulty. Today, even small carpentry firms rely on this technology. In Japanese factories, a multitude of monofunctional machines are connected in series to a multifunctional station. In contrast to European systems, multiple functions are often performed by a single machine. Both systems have advantages and disadvantages; however, the Japanese prefer theirs as it avoids the complete breakdown of the production step in case of a machine fault.

Modular Wood Unit Manufacturing

Although there is not yet a company in Europe that specifically focusses on wooden three-dimensional high-level components (units) construction, it is already being applied in Japan by Sekisui Heim, a company that delivers both steel frame units and customized wood unit houses (see Chapter 5, and in particular Section 5.4.13 for more details). Here, "two-by-four" and "two-by-six" construction methods are combined with the legendary "unit method" originally developed for steel frame units. All units are finished in the factory by fixing surfacing materials to the timber framing. Unit production processes are carried out on computer-controlled production lines and an advanced system called Heim Automated Parts Pickup System (HAPPS) supports all levels of enterprise resource planning. Work in the factory is done by both large-scale industrial machines and trained experts who carry out finishing tasks as well as rigorous and continuous quality inspection during production. Trained quality inspectors, in addition to IT-controlled monitoring using high-tech sensors, continuously inspect quality during and after each work process. The sooner a failure is detected, the easier and at lower cost it can be corrected. There are quality checklists for approximately 250 unique points. After the units have been built in the factory, they are delivered to the construction site and assembled. The frame is assembled and equipped with the prefabricated roof within nine hours in one day. From this point, it takes approximately 45 days for completion and handover. The closed frame structure is protected from exposure to wind, rain, and other unexpected circumstances and provides an optimized environment for fast and high-quality finishing.

Multilevel Integration in Timber Prefabrication

In every production process, different production levels must be integrated into a general production strategy that is based on a multitude of economic factors. Timber prefabrication, which already implements a high degree of CAD/CAM and continuous IT infrastructure, is an especially complex, interconnected set of sequential techniques, processes, and operations.

3D CNC Technology and Automatic Feeding

Many companies fabricating wood elements already implement CNC processing centres with fully automatic feeding. The computer-controlled organization of production in conjunction with the latest feeding technology enables the processing centres to be run with fewer staff while the product enjoys a higher level of customization and can be adjusted to specific use cases.

2.3.6 End-Effectors and Automated Processes

Wood machining is quite automated, and the use of different end-effectors in CNC machines is widespread in industrialized countries. Traditionally in manual wooden component manufacturing, each tool required a specific machine. There is a workstation for sawing, another station for shaping, and so on. Today, all these workstations have been linked and automated. Even though it could be difficult to saw, cut, plane, shape, mill, and drill in just one workstation, today, industrial robots can utilize different end-effectors that facilitate all facets of woodworking.

1. *Automated cutting and shaping.* Once either solid wood or engineered wood is produced, the first step to manufacture a component is to cut each element to an approximate size and shape, after which the element is usually machined (Csanády & Magoss, 2012). This step must occur before the element is machined. The accuracy of the approximation depends on the element type. For a structural element, 5 centimetres can be left before final machining. For furniture elements, this allowable variation may be less than 5 millimetres before its introduction into a machining CNC. A circular saw is proper for linear cutting. This saw

offers accuracy in linear shaping, and is rigid enough to maintain a straight cut. In manual cutting, circular saws were the cause of many injuries to carpenters' fingers. Band saws are flexible and better suited for cutting curved shapes. The band is continuous, normally turning in between two wheels, and the element to be cut is placed between these two wheels. The band is normally less than 5 centimetres wide. Working with a manual band saw can be quite dangerous. If the curve of the incision is too small, the band could break apart or even burst.

2. *Automated timber element machining*. The main purpose of wood machining is to create a special shape for an adequate posterior assembly. When joining two wooden elements without any other special accessory, a void or channel is created in one of the elements and a slot in the other one. The milling cutter is designed with a special shape. Tongue and groove, finger joint, and dovetailing are some of the most used wood joinery systems. Sometimes, glue, adhesives, or nails are needed to make the joint permanent. Lately, special steel connectors are replacing permanent wood joinery systems. New furniture or even structural elements are clear examples of applying this concept.

3. *Automated finishing*. After the machining of the wooden element, and normally before the assembly process, the finishing of the wooden element is done. Wood must be protected against humidity, solar rays, fungi, and others threats (when wood is covered, this may not be needed). First, to open the pores and smooth the surface of the wood, sanding is required. This process can be done by robotic end-effectors (Nagata et al., 2006). After this, the wood is impregnated with different chemical products. Painting and Varnishing must be done in special workstations; often closed cabins are used. To move from one workstation to another accurately, without inducing any scratches, robotic handling devices facilitate the transportation. To reduce time and space for wood machining, multipurpose CNC machines, where an element can be cut, milled, drilled, and sanded, are used.

4. *Tilting station (Figure 2.22c)*. Wall and gable elements are transported from tables to the tilting station. The lateral procedure of the tilting table is carried out on the chosen track, where the wall or gable elements are subsequently positioned into the appropriate magazines.

5. *Magazine for wall and gable elements (Figure 2.22a)*. All wall and gable elements are temporarily or permanently stored in magazines while the walls are being processed. The plastering work or the painting of elements, as well as the assembly of doors, windows, and shutters are completed here. The walls or gable elements can then be directly loaded onto either a freight vehicle or dispersing wagon.

6. *Nailing unit for the production of board batch elements (Figure 2.22b)*. The individual boards are loaded onto the nailing unit using a conveyor belt or cross conveyor. This can be done manually or automated. A lifting apparatus sets up the boards that are then pressed together during the nailing process. During the automated nailing process, a nailing image is rendered containing defined protected areas that simultaneously control the nail entry and depth. The length and width of the board batches are variable.

a) Magazine for wall and gable elements

b) Nailing unit

c) Tilting station

Figure 2.22. Automated work stations. (Photos: Weinmann Holzbausystemtechnik GmbH)

2.3.7 Factory Production Layouts

Timber panel construction has developed proper factory layouts for timber, in the sense that this type of component production often determines the final layout of the building. Factory layouts are typically designed for manufacturing throughout the whole process. Prefabricated house manufacturers in Germany mostly use the timber panel construction method because of its short assembly time. This method of construction does, however, need substantial investment in a variety of machines. These installations are identified as production centres that combine several production steps and production subsystems/cells. The end-products coming from these production centres are either single timber panels that are connected on the construction site or complete, fully functional, and equipped modules that only need to be delivered.

Prefabricated timber elements, because of their structure and weight, are advantageous in terms of logistics when compared with brick or concrete elements. Furthermore, the gradual integration of all processes, including logistics IT, is one of today's main goals of the precut timber industry. Gradually industrialized logistics networks are developing from centralized and hierarchical networks to highly flexible decentralized networks. If these networks are sufficiently integrated into the original equipment manufacturers (OEMs) IT structure, supply chain management could become the most important part in delivering on-demand customized prefabricated timber buildings. Through upcoming tracking technologies, for example, radio frequency identification (RFID) tags, the identification of a construction component together with advanced process control could additionally advance the industry. Figure 2.23 shows an exemplary timber element prefabrication layout.

2.3.8 Emerging Techniques in the Field of Timber Prefabrication

Through combined innovation in components (integrated with cabling, etc.), new modular building designs, new plug-and-play connectors and novel manufacturing systems (e.g., able to automatically insert cabling and plug-and-play connectors during the manufacturing process), the degree of prefabrication can be enhanced and the overall building construction time can be reduced significantly.

Up until now, the ceiling systems in wood construction have consisted primarily of wooden beams that become a planar building component with the plating of boards. The undersurface consists of wooden battens that, for the most part, are covered with either plaster or plasterboards on the construction site. The raw ceiling slab occurs after the erection of the walls and is covered with footfall sound insulation and floating floors with wet or dry treatment. A pipe system is inserted into the floor pavement to install the floor heating system. In the case of dry paving, the pipes are inserted into premilled fibre-boards; however, this type of pipe laying can be done only on-site. Therefore, a new ceiling system should be developed with the usage of fast connector technology (see also **Volume 1**, **Section 4.2** for more information on fast connector and plug-and-play approaches) that could be completely fabricated off-site.

Automated cabling in the SE of the factory (in combination with an adequate fast connector technology, see, for example, **Volume 1, Section 6.5.5**), which is more

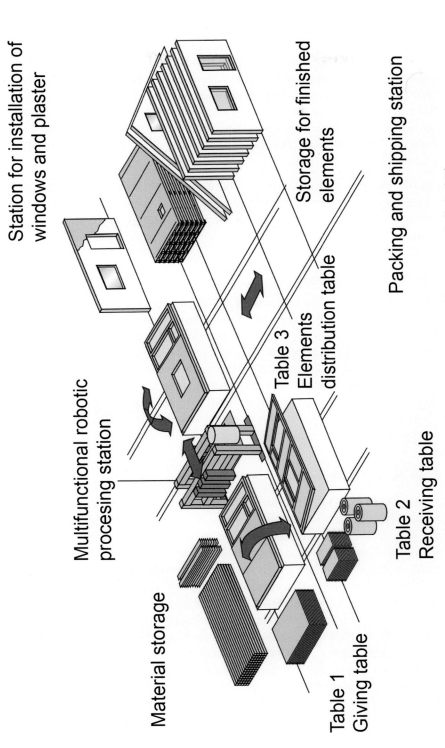

Station for installation of windows and plaster

Storage for finished elements

Packing and shipping station

Multifunctional robotic procesing station

Table 3
Elements distribution table

Material storage

Table 2
Receiving table

Table 1
Giving table

Figure 2.23. Exemplary timber element prefabrication layout. (Photo: Weinmann Holzbausystemtechnik GmbH)

or less directly connected to specific CAD data, is crucial for the fabrication of plug-and-play elements. The electric wiring in wooden prefabricated construction is up to date still completed primarily on-site. The following has to be considered:

- The installation of taut wires into fabricated walls requires great diligence so as to not damage the neighbouring cables and moisture barrier foil, as this damage could lead to condensation and a deficiency in wind proofing.
- Cable channelling, due to the taut wire channelling not being orthogonal and thus not in compliance with code (e.g., DIN 18015 in Germany), can result in problems/damages during the installation as the user is normally not aware of the positioning of the cables.

Because of these preconditions an installation principle should be developed that allows orthogonal cable channelling and in no place damages the moisture barrier foil while leaving enough space for the installation of plugs, cables, or rapid connectors.

2.3.9 End-of-Life Strategies

Timber recycling processes can be separated into two fields. First, the main form of the wooden component is somehow maintained, that is, there is a reuse of the wooden element. Usually, only structural elements can be reused by this technique and it requires cleaning, removal of nails, and protection against pathologies before relocating the element. This procedure is primarily done manually. Another method of recycling, perhaps allowing more automation, is to scrap, chip, or flake wooden elements and convert them into an engineered wooden element. If the quality of the source wood or vegetal source is not good enough for use as an engineered wood, the chips can instead be reused for creating combustible pellets.

2.4 Steel-Based Components

The production of steel requires manipulation techniques from the very beginning. Today, the whole process of creating steel products can be considered as medium to highly industrialized. Steel is an isotropic, manipulatable material, which can be used for linear elements, surface elements, and even other special shapes. Although the steel manufacturing industry is already quite an automated and robotic industry, the application of this material in construction is not always as highly developed as it could be. This section is an overview of steel component manufacturing. It focusses on the application of steel components in construction. In other words, the aim of this section is to create an overview and approach possibilities for automated steel component manufacturing. First, traditional steel element fabrication and manipulation techniques are reviewed. A detailed description of all metalworking processes would be far too lengthy; therefore, the processes are simplified to provide a clear overview.

2.4.1 History and Techniques of Steel Production

Iron and steel started to be used industrially and massively in construction during the 19th century. The very first furnaces for producing wrought iron were called bloomeries (Allen et al., 2009). During the early 19th century, modern blast furnaces improved the quality of output. The Bessemer process, patented in 1855, expedited the mass production of steel (Birch, 2005). Today, the steel industry mass produces standardized products that are used for railways, structures, building envelopes, and ships. Although they were invented at the beginning of the 20th century, it was not until after the Second World War that the use of electric arc furnaces was generalized (Russell et al., 2013). This technique permitted a lower production cost than that of an integrated steel mill that used blast furnaces. It is also more suitable for the reutilization of scrap. It was also after the World War II that the basic oxygen furnace steelmaking process was developed by Robert Durrer. As a result of the improvement of electric arc furnaces and basic oxygen furnaces between 1920 and 2000, labour requirements in the steel industry decreased by a factor of 1000, from more than three worker-hours per tonne to just 0.003. In both cases, continuous electrical supply must be guaranteed, meaning that a constant and reliable energy source is needed for production. Industrial steel production generally has some adverse effects on the surrounding area, and to avoid environmental costs, production must be enclosed.

2.4.2 Keys and Figures

The World Steel Association (WSA) states that in 2010, 71% of the world's steel production was accomplished using the basic oxygen furnace while 28% used electric arc furnaces. Until recently, only big industrial capitalists who controlled large markets could afford the steel production process. Today, the industrial production of steel (or industrial steelmaking) needs less of an investment compared to some years ago. Although this investment is still high, a 50-tonnes electric arc furnace is priced at around 7 million euros – much more affordable than in the past.

Today, steel production is concentrated in big companies. According to the WSA (World Steel Association, 2012), in 2012, companies with major steel production were ArcelorMittal, with 93.6 million tonnes of steel crude; Nippon Steel & Sumitomo Metal Corporation, with 47.9 million tonnes steel crude; and Hebei Group, with 42.8 million tonnes. WSA points out that in 2012, 1547.8 megatonnes were produced worldwide. The ranking of steel production by country is as follows. China widely outpaces all other countries, with production of 716.5 million tonnes steel crude. Japan and the United States follow with 107.2 and 88.6 million tonnes respectively. The WSA specifies that if "apparent steel use per capita" is considered, it can provide a glance of the kind of production type there is in each country (data are from 2011). South Korea has a consumption of 1156 kilograms per capita. It is followed by Taiwan with 784 kilograms per capita. In third position is Japan with 508 kilograms per capita. Construction consumes a larger proportion of steel in emerging countries. In those countries, construction covers up to the 50% of the total consumption. In advanced countries, construction equates to just about one third of total consumption.

2.4.3 Classification of Products

Steel is a massively produced and widely used alloy, mainly because it has very good properties for many purposes. The properties can be varied based on the final intended purpose of the steel components (Harvey, 1982). For instance, the tensile strength of steel is around 400×10^6 N/m² whereas that of pinewood is approximately 40×10^6 N/m². The modulus of elasticity (Young's modulus) is around 200×10^9 N/m². The density of steel is around 7850 kilograms per cubic metre whereas that of pinewood, for example, in comparison is only about 500 kilograms per cubic metre. This means that it is a relatively light material relative to its strength (as outlined in **Volume 1**, this is advantageous for automated/robotic processing and assembly of steel parts and components; in Chapter 5 in this volume it is also shown that the compactness and low weight of steel parts/components is also advantageous for automation in LSP). Steel can be easily machined, using thermal or mechanical techniques. Depending on its composition, steel starts to melt at approximately 1300°C. This must be considered in construction and fireproof solutions must be arranged to avoid fast melting and collapse of the structure (Campbell, 2011). Steel is a highly conductive material, which means that it has to be protected against electrical discharge from lightning or other sources. Because of this, steel can also be manipulated with controlled electrical discharges, permitting the machining of elements. Corrosion is an electrochemical oxidation of metals that can often affect steel. This must be avoided, as it degrades the structural, formal, and permeable properties of steel. Interestingly, corrosion itself, if properly applied, can protect the integrity of the steel members. This is the case of "weathering steel" (Thomas, 1995), better known as COR-TEN steel. Steel must be combined properly with other materials, especially if it will experience water contact. Varying steel properties can be either profitable or problematic, depending on the situation. The characteristics of steel must always be under control, and therefore steel requires constant maintenance, especially if the material is exposed. As we will see in the following sections, the production of steel involves applying protection, machining elements, and finally assembling those elements. Different types of steel can be manufactured according to the material composition, shape, form, and function of the element.

Material Types

Steel can comprise various components, leading to different steel types and uses. The European Standard EN 10020 **(**EN 10020:2000-07**)** classifies steel grades defined in the following ways. To briefly explain that classification, the primary categories of steel are considered as non-alloy steel, alloy steel, and stainless steel. According to the EN 10020, even though every steel is actually an alloy, a non-alloy steel is produced when carbon makes up less than 2% of the total composition. Alloy steels include some other component, such as manganese, nickel, chromium, molybdenum, vanadium, silicon, or boron. In most cases, these compositions do not exceed 8%. Finally, steel is considered stainless in compositions containing at least 10.5% chromium and less than 1.2% carbon. The EN 10020 also classifies steel according to its future purpose. Steel production varies greatly when it will be used as a cutting tool compared to if it is used for cladding or in outside ambient conditions.

Steel Elements in Construction According to Shape

If we focus on the steel products that are used in construction industry, elements can be grouped according to the initial shape. The shape determines the future manipulation process of the steel product. Even more, it defines the function that the product will fulfil within the building. The principal element types in steel construction are generally linear and superficial. Besides these elements, there are also some special elements. The production of all elements is based on mass production, and is explained further in the next section on how those elements are produced.

1. *One-dimensional, linear elements*. Included within this group are steel profiles, tubes, and cables. Some of these elements are used for structural purposes. There is a long tradition in the use of steel structures. The profile of these types of steel geometries is already predefined. Structural calculations have been developed and simplified for such cases. Besides structural elements, tubular profiles and cables are quite common for services and installation. Plumbing is a clear example of this. They can also be used for the substructure of partition elements. There are also good architectural examples of using steel as a substructure for curtain walls, windows, or even doors. Reinforced concrete also uses linear steel elements; however, this falls outside the scope of this section.

2. *Two-dimensional elements*. This group of elements is composed primarily of steel sheets, which are used for cladding or internal distribution. The thickness of these sheets in this case is not larger than 5 millimetres. On the other hand, steel sheets can also be used for structural purposes, in special or singular buildings. In these cases, the steel sheet is typically thicker and can be up to 60 millimetres thick. This kind of structure is more traditional in the shipbuilding industry, where the vessels are formed by thick steel sheets. There are, however, examples outside of the shipbuilding industry also. In civil construction, this kind of material is used in beam bridges to create a structural support that is not easily obtained by ordinary profiles.

3. *Three-dimensional, special shapes*. These types of elements are mainly produced by casting methods. They can be used for special purposes, but there are also common elements such as valves, water taps, and other fixtures in the plumbing industry. Handles, hinges, and many furniture elements are also special steel shapes. Some special shapes in the form of structural joints and distribution systems also exist, but these can be considered residuals.

2.4.4 Manufacturing Methods: Steel Elements

Before the implementation of a steel component in construction, four main steps must be completed. First there is steel basic element production, which is considered as production process, in which no customization is achieved. Next, each element is cut, machined, and bent according to the required function and form of the component. These machined elements need to be protected against hazards such as corrosion. Finally, those machined and protected elements are assembled (Waters, 2002) into components. The ability to melt the material also provides the opportunity for moulding the material. The initial method was to poor melted steel into a simple mould. Today very complex and sometimes even flexible, robotic moulds and advanced injection systems are used.

Besides moulding, forging, lamination, and extrusion are used to to generate the desired basic elements. Those processes are metal shaping procedures that require compression forces. The internal substructure of steel is compressed and the inner strength of the material is ameliorated. These processes are performed by penetrating steel, cold or hot, into a "die" with large pressures. The die can be open or closed. The extrusion process for steel profiles was patented by Joseph Bramah in 1797. In 1849, Alphonse Halbou patented a method for laminating hot extruded profiles using a rolling system or a laminating line. It consists of rolling steel into one or several dies until it acquires its final shape (Bauser et al., 2006). The basic forging process consists of hammering a steel element using the strength and ability of the blacksmith. Today, forging consists of inserting a steel sheet into a two piece mould or "close dies". These moulds are forced together with a press. As outlined in **Volume 1**, **Section 4.4.4**, processes as molding and stamping can be considered as rather parallel processes that allow for a fast mass production of the basic steel elements/parts. Processes as CNC controlled cutting (laser, water jet, etc.) or additive manufacturing (3D printing) can be considered as rather serial/sequential processes that allow for direct customization of the basic element/part but are not as fast and productive as parallel processes. Extrusion systems are able to combine elements of parallel and serial/sequential processing.

Steel Element Bending and Machining Processes: The Tools

Mechanical tools can be used for steel element machining or manipulation (Maekawa et al., 2000). What do we mean by steel element manipulation? The steel is manipulated to remove material from the rough element. It is then bent and torched to achieve the desired form. These tools are used for cutting out a piece from the steel element. There are mechanical, thermal, and abrasive techniques to do this. It can be stated that in these processes, there is customization of the mass-produced element, that is, the steel profiles and sheets, as seen in this section. In the following section, it will be seen how these tools have been adapted as CNC controlled manipulators and end-effectors.

1. *Mechanical techniques* are used when the steel element is cold. These are based on the traditional method of steel machining that requires mechanical tools. Bending (Wanheim, 2004) can be used for any of the mass-produced metal sheets, profiles, or tubes. In the past, it was a manual technique. Bending is very useful in the construction of cladding systems. Structural profiles can also be bent to create special shapes. In the service sector, the plumbing industry needs to bend its steel and metallic tubes. Drilling can be considered simply as making holes in a steel element. The dimension of the hole is determined by the diameter of the tool, that is, the drill. The drill simply penetrates the material. The elements generally remain still on a bench (the bench can move), and the drill turns (Knight, 2005). Milling consists of removing material from an element using a milling cutter. Unlike drilling, the milling cutter is moved to create a previously specified shape. The shape of the element will have some limits. For instance, if an inner cavity is removed, the cavity will have a minimum radius of the milling cutter. In turning, the element that is going to be mechanized turns around an axis. The cutting tool shapes the steel element via this rotation. The output element's shape is cylindrical and very accurate. This type of element is not often used in construction.

2. *Thermal techniques* involve the melting property of steel. Heat, applied on a focalized surface, can be used to make holes (Jeffus, 2011). For instance, oxy-fuel cutting is a process that uses fuel gases and oxygen to apply heat on the steel surface and melt it to produce a cavity in it. It is used when the cutting element has a considerable thickness. Alternatively, arc cutters involve use of electrodes. They were first created at the beginning of the 19th century and have since been developed into plasma arc cutters. Arc cutting is probably the most used thermal technique (Jeffus, 2011). An electric arc high-voltage sparks to ionize the air through the torch head and initiate an arc. A gas purifies the melted element and leaves it without any gas inside. This means no bubble is created. The plasma melts the metallic element. This plasma rapidly removes the material, and a precise cut is made. Finally, the laser cutting technique was developed with the first CNC systems (Caristan, 2004). A high-power laser output is directed at the steel element to remove the material. The material is melted, vaporized, or blown away.

3. *Abrasive techniques* are based on abrasion, which induces erosion of the material. They are useful for achieving a high-precision element or producing very fine finishes with high precision. Grinding is basically an abrasive method. Normally, these techniques use an abrasive wheel, made of diamond or some other organic material (Knight & Boothroyd, 2005). Like in turning, the required precision in construction is not very high, which means this technique is not widely used. Electrical discharge machining, also known as EDM, is a technique that was developed in the mid-20th century. Fast current discharges between two electrodes remove the material from the steel element. The element is dipped in a dielectric liquid to facilitate the process. A wire or graphite electrode is used to create the erosion. The shape of the graphite electrode defines the mechanization of the steel element; it works like a kind of negative mould. Water cutter jets apply pressurized water, mixed with an abrasive powder. The abrasive powder is usually made of granite.

Steel Element Protection Processes

Steel must be protected from environmental hazards. The design of buildings must take into account these protection systems. The main problems in steel construction are corrosion, fire, and electrical discharges created by lightning. Every country or region has its own specific regulations against fireproof or anticorrosion systems. It is common, for instance, that fireproof regulations define a minimum structural strength during evacuation in case of fire. Galvanization is used to improve the properties of the steel elements surface and avoid corrosion. Normally, this process consists of dipping the steel element in a molten zinc bath (Maas & Peissker, 2011). Once the steel element is assembled on-site, it can be coated with special painting for protection from corrosion and fire. Weathering steel is a well-known process in steel construction. It is based on forming stabilized rust that protects the steel element from further rusting (Thomas, 1995), which can be done by hand, pouring chemical products directly onto the steel element. In a building, steel made structures are often covered with plaster boards. This is generally a manual procedure. Sprayed fire protection consists of projecting perlite, vermiculite, and Rockwool onto the structure. It is projected after the steel structure has already been built. Steel polishing is used to remove material and obtain a smoother surface. Besides imparting a shine aspect, this prevents corrosion.

Steel Element Assembly

Once the elements are mechanized and protected, they can be assembled using a variety of different methods. Riveting is the oldest method for assembling iron or steel elements. A cylinder with a head on the top (could be considered as a big nail) is introduced in the holes of the two elements to be fastened. It requires the previous drilling of the steel element (Purohit & Sharma, 2002). This kind of assembly imposes a permanent joint, meaning that the union must be broken if disassembly of the elements is required. Rivets can be of many sizes and offer different union strengths. For the bolt and nut system, as in the previous method, the steel element requires a previous mechanization, which is making a hole in both elements. Depending on the strength required, the quantity and quality of the bolt and nut will differ. These elements can always be unscrewed and disassembled, not only in the structure, but also in the distributing system. Normally, a transition plate between the joined elements guarantees proper fixation. The thermal techniques for steel assembly are similar to those in the section on steel element bending and machining, that is, oxy-fuel welding, arc welding, plasma arc welding, and laser welding (see also Purohit & Sharma, 2002).

2.4.5 Possibilities for Industrial Customization

Steel is probably one of the materials requiring the most intense, resource and energy intensive manipulation in the manufacturing of the basic element – however, (as outlined previously) due to its compactness and low weight steel elements are more advantageous than other basic elements when it comes to assembling this basic elements (e.g., by automated systems or robots) into components. Steel must first be melted to create the basic material, then extruded or casted to create the standard element. After this, the element can be bent, machined, and protected with special coatings. Specialized software and industrial robots offer great possibilities for a customized process. Until some years ago, if a steel element were to be machined in a CNC, the geometrical information of the element had to be manually inserted into the CNC. Every element of the building needed to be machined this way. The CAD or BIM, CAM, and CNC coordination facilitates the link between the computer drawings and designs and the industrial robot (Altintas, 2012). Today, once a steel structure is designed, all of the elements could be machined automatically in the CNC. The CNC will machine the elements according to the CAD or BIM drawing. The software separates all elements of a whole building 3D model and links the information with the machining centre. Does this mean that the whole building, including all details, must be drawn for every project? Not necessarily. If standard steel profiles and machining types of the elements are used, there exist many standardized solutions for those cases. Detail libraries can be created, the details of which are composed based on the project's geometry. Sometimes, a link is needed between the CAD and the CNC. The diversity of BIM and CAD developers as well as CNC producers sometimes makes the coordination between them quite difficult. A simple plug-in can usually resolve these problems.

This coordination drastically reduces programming time of the CNC and eliminates rework. Tekla and Revit, for instance, are two pieces of specialized software for steel structure design that work using linear profiles. Alternatively, software such as Solid Edge and Solid Works are more specialized in creating in new steel shapes. These programs have special tasks or tools for each manipulation process such as bending, milling, and so on to "control" that the designed element can be realistically produced with existing techniques. In the aerospace and ship industries, interesting software such as Catia helps facilitate the gap between design and production (Góngora, 2013).

This kind of software is also useful for product lifecycle management. They offer the possibility to draw all of the elements with accurate detail. It can also be useful for structural calculations. Controlling the whole process and designing a product that is reliable can be done faster thanks to this special software. For instance, Lantek's (2014) software creates and controls the integration of many processes such as project setup, budget, financial planning, purchases, production, staging, certification, invoicing, reporting, and closing. Software and industrial robots are becoming more automated. This means that less control over the process is needed and elements can be manipulated automatically. It can be said that, this way, the production doesn't depend on standards. There can be a customization of the final product, but not necessarily a tailor-made solution. The solution will depend on the degree of coordination between the software and the industrial robot so as not to spend too much time on the process.

2.4.6 End-Effectors and Automated Processes

The previously explained manual techniques for protecting, machining, and assembling steel elements are also used in industrial robots. The industrial robot offers a faster and more reliable mechanization (Altintas, 2012). Some techniques, such as electrical discharge machining and laser cutting, were almost created as end-effectors, or at least as part of a numerical control.

Automated Forging, Bending, and Machining

Automated forging is mainly used in the automotive industry, to facilitate massively produced elements. For instance, this is the case of the chassis (Loire Safe, 2014). These are industrial robots designed with a defined purpose. This case is not considered as just an end-effector. The entire machine is designed for a purpose. Its purpose is well defined and the performance is specifically directed towards a set goal.

Sometimes, an external robot can help with part of the process to increase productivity and security. In construction, the products produced by automated forging are mainly used in services and installation such as water sinks, heating devices, and so on. For instance, in automated bending, today a press system is used (Danobat, 2014). The steel element is typically moved to be folded several times. These presses work with shear or with cylindrical presses, depending on the kind of shape required. The shear can produce smaller angles while the cylinder is more appropriate for more open curves. The shipbuilding industry also uses large presses for manufacturing the curved steel sheets of the vessel. In these cases, though, there is no die and no mould. The steel sheet must be folded in a different way every time to achieve the required form.

For cutting, drilling, and removing material, end-effectors in industrial robots have become more prominent in the car industry (Altintas, 2012). The industrial robot technology using the manipulation and mechanization of the structural elements can be completely automated. Mechanical drilling requires a very stable machine or robot. Some of the CNC machines have many end-effectors, so the steel element can be machined with different techniques without moving the element to another machine.

Automated Steel Protection

Painting and galvanizing are processes that are held in a special line. The processes have several stages and the elements are typically transported using overhead conveyors. The surface treatment tunnel (Geinsa, 2014) gives specific characteristics, such as increasing the resistance to corrosion and oxidation (number of hours salt spray resistance) to the steel element and must be subjected to a series of chemical and washing processes. Cataphoresis is a dip-coating process that is completely automated and based on charging the steel element to a continuous negative electrical voltage and the bath to a positive electrical voltage, attracting the paint particles and obtaining a uniform paint film and outstanding finish with corrosion protection (Geinsa, 2014). There are also painting robots. Once the surface is ready, a robot can paint the required surface. The appropriate type of robot to use will depend on the range of product. After painting, drying ovens are used to improve the drying process. Less time is required to get the surface dry enough for the next step. Polishing robots (Autopullit, 2014) can be used for complex elements and surfaces.

Automated and Robotic Steel Assembly of Components

Steel construction elements can typically be assembled in one of two ways: on-site and off-site. Robotic steel component assembly is highly developed in the automotive, aerospace, and shipbuilding industries. In construction, it is developed only in certain countries, including both Japan and Korea. These are explained further in the next sections.

2.4.7 Factory Production Layouts

An automated factory that agglutinates steel element production, mechanization, and assembly can be difficult to find. The processes involved from steel production to final assembly can be quite different across various factories (Danobat, 2014). First, the steel profile and sheet are mass-produced in a factory. Then, steel element manipulation is often done in another factory. Finally, element assembly can be done either in a factory or on-site. When considering steel element machining, there are machine-robot builders that offer a complete factory layout for the entire machining process. Production lines offer an automated storage of the product for different kinds of profiles. After this comes automated profile cutting and drilling.

a) LaserCUSING® machine solution

b) LaserCUSING® is not just used with steel; gold elements can be produced too.

Figure 2.24. Additive manufacturing of metallic elements. (Photos: Concept Laser)

This is followed by a series of automated welding processes. Finally, non-destructive inspection of the machined elements is completed using ultrasonic waves.

2.4.8 Emerging Techniques in the Field

Apart from traditional end-effectors, CAD/CAM coordination opens up possibilities for new techniques, meaning the adaptation and development of CAD/CNC technologies provide new fields for creating and implementing new technologies.

This is also the case with 3D printers. Although these printers were mainly used with other materials, some companies, such as Concept Laser, have developed an interesting application for producing complex metallic elements and so called direct components (components that can be used without the necessity of further stamping or hardening processes) directly as functional elements in systems or components of products. The LaserCUSING® (Concept-Laser, 2014; Figure 2.24) process is used to create mechanically and thermally stable metallic components with high precision. Depending on the application, it can be used for stainless and tool steels, aluminium and titanium alloys, nickel-based super alloys, cobalt–chromium alloys, or precious metals such as gold or silver alloys. With LaserCUSING®, finely pulverized metal is fused using a high-energy laser. After cooling, the material solidifies. Component contour is achieved by directing the laser beam with a mirror deflection unit (scanner). Construction takes place layer by layer (with each layer measuring 1 to 200 microns) by lowering the bottom surface of the construction space and applying and fusing more powder. Concept-Laser systems stand out because of their stochastic control of the slice segments (also referred to as "islands"), which are processed successively. The patented process significantly reduces tension during the manufacture of very large components. An effective solution for working with big steel elements, where the working station could also require large industrial robots, is the flexible drilling head developed by Mtorres (2014). It is a kind of climbing robot that drills in complex surfaces, for example, in the fuselage of aircrafts. The robot is able to recognize the exact location in which it is supposed to drill.

2.4.9 End-of-Life Strategies

As previously stated, steel elements must be protected properly to reduce the risk of corrosion. This means that steel elements/components have to be periodically

Figure 2.25. Steel wind blade cleaning robot RIWEA developed by Fraunhofer IFF. (Elkmann et al., 2010)

maintained. To develop an automated form of maintenance, measurement and data scanning play a crucial role. Furthermore, if the steel element/component is not functional anymore and has to be abandoned, there is still a solution for recovering the material. This is called scrap recycling.

Automated Measuring and Maintenance

Steel requires maintenance for continuous adequate performance. This maintenance can be difficult and expensive work. Therefore, some automated solutions for this maintenance have been developed. First, an automated inspection is needed to measure the existing situation and possible defects of each element/component. If damage exists, the element may lose its geometrical properties. Laser scanning technology enables the detection of accurate geometry of the steel element (Bosche & Haas, 2008). Once an accurate diagnosis is made, a steel maintenance robot can operate in the affected areas. The sectors most experienced in this area are probably bridge and wind turbine cleaning and repairing, as is the case of the RIWEA robot developed by Fraunhofer IFF (Elkmann et al., 2010; Figure 2.25).

Automated Scrap Recycling

Through melting processes, steel elements (in case the components were properly designed for disassembly back into mono-material basic elements; see also **Volume 1, Section 6.1.3**) can be easily recycled. According to the World Steel Association, around 30% of the steel produced in the world comes from steel scrap. Scrap recycling consumes less material and energy than obtaining steel from minerals. In addition, it creates less air pollution than traditional processes. Can steel be considered a totally recyclable material? To do this, some precautions must be taken. There must be control of scrap quality, especially with the previous detection of radioact-ive material. Advanced countries employ a high level of control over the quality of scrap steel. There are automatic radioactive detection systems. Steel elements are present in our everyday lives. In other words, there are numerous places where steel elements can be found. Because of this, an automated steel collection system is dif-ficult to design. To avoid illegal and dangerous situations, scrap-collecting systems should be regulated and rigorously tested.

Building Module Manufacturing

This chapter provides examples of the manufacturing of medium/high-level building blocks (in the following referred to as modules; building modules, prefabricated bath modules, or assistance modules that can also be referred to as building subsystems). Those medium/high-level building modules are in hierarchical, modular building product structures independent building blocks (see **Volume 1**, **Section 4.2**) that are delivered by the companies that produce them (Tier-1 suppliers) to original equipment manufacturers (OEMs; for a detailed explanation of the OEM-model, see **Volume 1, Section 4.3.3**) as, for example, large-scale prefabrication (LSP) companies (see in particular Section 5.2.6 in this volume), conventional construction sites or automated/robotic on-site factories (see **Volume 4**). LSP companies such as Sekisui Heim, Sekisui House, Toyota Home, and Misawa Homes (Hybrid), in particular, have altered the building structures, manufacturing processes, and organizational structures dramatically in comparison to conventional construction in order to be able to assemble in their factories and on their final assembly lines as many prefabricated medium/high-level components from Tier-1 building module suppliers as possible. As outlined indetail in **Volume 1, Section 6.3.7**, this practice can be considered an ROD method that reduces the amount and variety of assembly activities on the assembly line and thus helps to create a SE in the LSP factory or automated/robotic on-site factory. In particular, in Japan, both prefabrication companies (OEMs) and module manufactureres (Tier-1 suppliers) are integrated in a well established OEM-like industry organization. Toto and Inax (both major suppliers of bath and kitchen equipment in Japan), for example, prefabricate plumbing units (especially bath cells) and deliver them as Tier 1 suppliers to LSP copanies, contractors and automated/robotic on-site factories. The use of prefabricated plumbing modules in the form of completely equipped three-dimensional cells is widespread in Japan.

The industrial manufacturing of bathroom and kitchen modules began in approximately 1920s in the Scandinavian countries (Sweden, Finland, and Norway). Long winters and short summers, and therefore short periods in which final on-site construction was possible, necessitated the use of prefabricated elements. In Germany, the industrial production of rapidly installable, three-dimensional subsystems/modules became popular in the early 1960s. The creation of plumbed rooms on construction sites in a conventional way is still often a bottleneck, as many trades

a) Production of waterproof
plastic bottom plates

b) Equipping of bottom plates
with technical installation

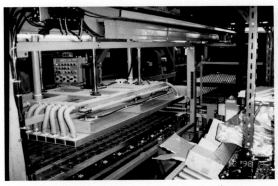

c) Automated production of interior finishing

d) Assembly of (sandwich) wall panels

Figure 3.1. Toto assembles customized, prefabricated bath modules for OEMs on the production line. (Photos: Taken at Toto Ltd. factory)

must be coordinated simultaneously, often in a limited space to carry out precision work. The prefabrication of modules, however, can take place off-site in an SE and in parallel to the actual on-site construction, without logistical difficulties and disturbance of the actual construction process. In their early development, prefabricated subsystems were commonly used in the construction of hotels, hospitals, and nursing homes, and especially in Japan for construction of high-rise buildings.

Today the use of prefabricated modules is no longer limited to a specific construction type, and bath or kitchen rooms based on high-end prefabricated plumbing modules (e.g, manufactured by companies as Hitachi, Toto, Inax, Deba, Lixil) can hardly be distinguished from a conventionally built bath or kitchen rooms. Prefabricated units from the company Deba, are, for example, also used as building modules for large passenger ships. Three-dimensional prefabricated plumbing modules can be delivered as compact modular units or as knock-down kits. In Japan, in many cases plumbing modules are similar as the units for prefabricated buildings (see, e.g., Section 5.4.10) prefabricated on a production line where carrier elements (e.g., the bottom part of the modules serving as a kind of chassis or template) are equipped with different finishes, bath equipment, and appliances to suit individual needs (Figure 3.1). As plumbing modules are high-level components and reduce the

e) Automated production of finishing elements for the bath modules

Figure 3.1 (*continued*)

amount of parts and work activities in downstream processes, their use in respect to an intended automation/robotization in the downstream value added steps carried out by the OEM is advantageous.

Plumbing modules are demanded especially by all major LSP companies (in particular Japanese LSP companies, see Chapter 5) that fully integrate equipped bath cells as subassemblies in the factory (Sekisui Heim, Toyota Home, Misawa

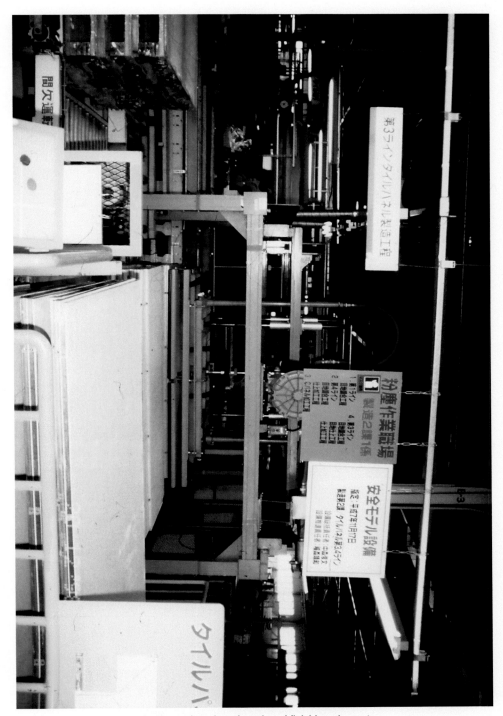

f) Automated sorting of produced finishing elements

Figure 3.1 (*continued*)

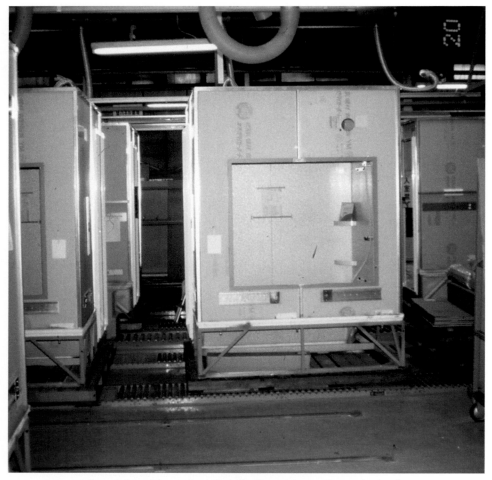

g) Completion of three-dimensional bath modules on the production lines

h) Finished modules

Figure 3.1 (*continued*)

Homes Hybrid) or on the site (Sekisui House, Daiwa House, Misawa, Mitsui). In the case of Sekisui Heim in particular, the stringent organization along the production line supports an OEM-like integration structure, as it allows the material supply to be entered directly into the final assembly and sequenced and clocked over dedicated spaces and gates along the production line.

4 Comparison of Large-Scale Building Manufacturing in Different Countries

Chapters 4 and 5 deal with the off-site manufacturing of complete buildings composed of low-level components, medium-level components, and very high-level components (units) and focus in particular on systems and kits produced using automation and robot technology in larger quantities (large scale).

In this chapter, four countries are analysed: Germany, the United Kingdom, Spain, and China. The proportion of large-scale building system manufacturing within the building industry differs from one country to another, for various reasons. Technological background, existing inventory of buildings, or educational level may provide clues about how the building manufacturing industry has been developed. The degree of automation of manufacturing processes in each country also plays a strong role. What are the reasons of such a high level of automation in manufacturing processes? Each country has different possibilities and constraints to define policies and promote large-scale prefabrication (LSP). For instance, if a country must employ a low-wage workforce because of its socioeconomic background, this automated production model may not be the most suitable for its economy.

If we focus on technological background, Japan has a very long tradition in automated production processes. It is a step ahead of the rest of the countries regarding LSP. For this reason, Japan is considered separately in Chapter 5. LSP has also been quite developed in Germany, and accordingly, in Germany, many interesting companies prefabricate a considerable number of buildings per year, though in a less automated way when compared to Japan. In countries with less tradition in automated building manufacturing, such as Spain, the United Kingdom, or China, it is not so common to manufacture complete buildings in large quantities. Because of lower technological backgrounds in the building industry and the still very low wages in the conventional construction industry, traditional building processes cover a larger proportion. Even though countries such as Spain and China have experienced a large construction boom over the last decades, LSP has not been developed so deeply yet. However, in particular in China, the cooperation of Vanke and Broad holds strong potential for systematization, and perhaps later on, automation of the production of building systems. The Korean case is also promising in this regard. The technological background has been constantly developing and the long tradition in the highly automated shipbuilding industry offers new possibilities to perform LSP.

The subsection on every country starts with a historical and general overview of the LSP field. To gain an objective perspective, data of companies, in each country, are compared (e.g., year founded, business volume, output quantity, number employees, range of products, price per square metre, and number of factories). After this, the sections focus on logistics systems, manufacturing methods, and the composition of building systems. Furthermore, the business strategies of the companies and future developments are presented.

4.1 Germany

Building prefabrication in Germany can be classified into wood-based prefabrication (prefabrication of small- and large-scale wooden frames/panels) and steel-based prefabrication (prefabrication of three-dimensional space frames). Further, the prefabrication of concrete and brickwork panels plays an increasingly important role. In contrast to Japan, steel is not accepted as a building material for housing in Germany, and the steel-based prefabrication industry mainly manufactures functional buildings (hotels, laboratories, hospitals, homes for the elderly, offices, factories). The larger prefabrication companies have a business volume (turnover) maximum up to €200 million per year, and manufacture at most 1000 buildings per year and generally have no more than around 800 employees. Research showed that most German prefabrication companies are relatively old (founded between 1900 and 1950) and, in contrast to Japanese prefabrication companies (which transitioned from traditional wood-based construction to radically new product structures faster due to their young age and their roots in chemicals, electronics and automotive industries, see Section 5.1.5), did not considerably change the structure of their products. Manufacturers such as Weber Haus and Baufritz achieve a high ratio of automation and excellent quality control in their factories; however, the industry is (owing to a lack of concentration on large companies and relatively low output by individual companies) only in the beginning phase of developing the original equipment manufacturer (OEM)-like integration structure on which the Japanese prefabrication industry is already based (see Section 5.2.6).

4.1.1 Wood-Based Housing Prefabrication in Germany

The share of prefabrication in the housing industry is rising slowly but continuously. Currently, its share in Germany is approximately 15 to 20% but varies with geographical location. In general, the share of prefabricated buildings is higher in southern Germany than in the north. Reasons for this are the economic conditions and low unemployment rate (typically not more than 3 to 4% around Munich and Stuttgart), leading to a high number of privately owned homes and thus also to a higher amount of construction activity. Further, in the southwest (Schwaben), the percentage of owner-occupied housing is traditionally higher and people tend to spend more money on home construction. On average, about 20,000 prefabricated homes are constructed per year through three types of prefabrication:

- Type 1: factory-based wood panel prefabrication
- Type 2: massive factory-based constructions (brickwork panels, concrete panels)
- Type 3: carpenter-like prefabrication

Table 4.1. *German large-scale builders based on wood*

Company	Since	Business volume/ year[a]	Output/ year	Employees	Range of products	Price range[b]	Number of factories
Baufritz GmbH & Co. KG	1896	59	152	240	Wood-panel construction with mainly large panels	From 2000	1
Huf Haus GmbH & Co. KG	1912	83	100	280	Wood-frame construction, small and large wood panels	From 1800	1
Bien-Zenker AG	1902	128	800	580	Wood-frame construction	From 800	1
KAMPA Haus GmbH	1900	190	1,086	800	Mixture of wood-panel and wood-frame construction	From 600	1
Weber Haus GmbH & Co. KG	1960	158	700	930	Wood-panel construction with mainly large panels	From 1200	2

[a] In million euro.
[b] In euro per square metre.

Table 4.1 gives an overview over the major German players in the field of wooden prefabricated houses and their performance. Currently in categories 1 and 2, 30 to 40 companies are competing in Germany (some of them are also involved in type 3). The companies with the highest yearly output of prefabricated homes are Kampa House and Weber House. During the last decade, Kampa House has incorporated many companies and brands such as Hebel Haus, ExNorm, Libella, NovyHaus (Austria), Casa Libella (Italy), Trend Haus (Hungary), and Kampa-Budizol (Poland) and thus managed to cross the 1000 houses per year margin. The yearly output of most other prefabrication companies lies between 100 and 600 houses per year. Compared to the production volume of Japanese prefabrication companies (smaller companies such as Toyota Home: 5000 houses per year; larger companies such as Sekisui House and Daiwa House up to 50,000 houses per year), the output of buildings per year is rather low. This also explains the – in comparison to Japanese companies – low degree of automation and lack of factory organization based on continuous uninterrupted material flow or production line focused manufacturing layouts. The low yearly output prevents investments into further automation, expensive robot technology, and new organizational arrangements. However, future fusions or cooperation between companies might allow maintaining the economies of scope (through offering buildings through different brands) while improving economies of scale and thus positively influence further investment ability into automation (and ultimately trigger a performance multiplication effect [PME], see also **Volume 1**). Furthermore, current attempts by many companies to increase the yearly output of houses by expansion to Scandinavian countries and Eastern Europe might increase manufacturing volumes and thus the ability to reinvest in systematization and automation. Currently exports account for less than 20% of the output of major companies but a potential and intended increase of up to 40% is discussed within the industry.

The Manufacturing Process

Lead times and guarantees: In general, companies are able to plan and deliver buildings within 8 to 10 months. The guarantee for the building (structure) is generally 30 years and the guarantee on technical equipment 2 to 10 years. The guarantee is interpreted here in a rather European or German sense: in case of obvious damages or breakdown the company is informed by the user and fixes the issue. In Japan, guarantee (in many cases also 30 years or more) is interpreted as a guarantee that damages will not occur, that technical systems do not break down, and that the building and all of its functions can be used without interruption – therefore Japanese companies proactively inspect buildings (see Section 5.5.1).

Product (panelized wood construction): Major prefabrication companies rely on panelized wood construction. Panels are built up by a post and beam structure (wooden frame), which is filled with insulation and, in many cases, equipped with an additional layer of insulation on top of this. To fulfil energy consumption standards, outside wall panels are today, in many cases, thicker than 40 centimetres and the tendency leads to even thicker panels. Panels can be small, medium, or large (e.g., as large as one side of a building). The panels can be completely finished in the factory (windows, outside facade, and paint inside). In general, there is an increasing trend towards large panels that are fully finished in the factory (e.g., Baufritz, Huf Haus) to reduce task variability and amount of assembly operations on-site to a minimum.

Off-site manufacturing: The cutting, bolting, and sequencing of post and beam elements are automated in the factories of large companies (in most cases with systems for a direct translation of computer-aided design [CAD] data into computerized numerical control [CNC] of machines). Furthermore, nailing bridges and automatic turning or butterfly tables are a standard equipment. Some companies (e.g., Baufritz) have already automated the setup of the posts and beams (wooden frame) and their nailing to a top plate through the use of robotic fixtures and jigs. The inlay of insulation into the wooden frame on an assembly and infill tables is done in most cases manually and by assistance through mechanised equipment as, for example, balancers. Overhead manipulators (OMs), balancers and other material handling or logistics systems assist the transport of the parts as well as panels between different workstations. Furthermore, multifunctional bridges (for the automation of tasks as nailing, cutting, grinding, part/component positioning, etc.) are used to process panels, and butterfly turning tables support the orientation of the heavy panels in the final stages. Magazines in which the panels can be placed vertically are used to store elements between workstations, process steps, or after completion. All in all, the material flow can be characterized as flow-line–like (with many elements of workshop-like organization) but in contrast to Japanese companies, is not as chain-like or production line–like.

On-site assembly: On the construction site, buildings are assembled together from a multitude of panels. Depending on the size of the building and panels, this process takes one to two weeks. After this, the building is weather proof and finishing work can be done, which usually takes another couple of weeks. In comparison, Japanese prefabrication companies are able to close the building envelope between one (e.g., Toyota Home, Sekisui Heim) and several days (e.g., Sekisui House, Daiwa House). It can be said that in German prefabrication – although the majority of manufacturing operations are shifted to a structured environment (SE) in the factory – the individuality of the building is currently generated mainly by workshop-like organization at individual stations. By contrast, in Japanese prefabrication, variation and individualization are generated by modularization and one piece flow in a flowline or production-line–based organizational setting. However, this requires that the product structure of a conventional crafts-based product "house" had to be changed. The wooden frame structure of prefabricated buildings in Germany still strongly corresponds with the structure of conventionally built buildings and a significant alteration of the product towards being more prefabrication and automation friendly (in terms of modularization or robot-oriented design [ROD]) has not yet occurred.

It is characteristic for the industry that, except Baufritz and Huf Haus, companies focus on the lower end housing market and thus on outperforming conventional construction by offering less expensive solutions. Baufritz adds value by combining design with architectural quality with ecological aspects and exceptional quality of building materials. Huf Haus adds value through modern and exclusive design along with high-quality building materials. Most other companies make compromises in terms of design, ecological performance, and quality of building material to be able to offer as many square metres for as low cost as possible.

4.1.2 Steel-Based Building Prefabrication in Germany

Kleusberg, Alho, and Cadolto are all well-established companies with long-standing traditions that have evolved from carpentry and construction companies (Kleusberg, founded 1948; Alho, founded 1967; Cadolto, founded 1890). Kleusberg owns two factories (Wissen and Kabelsketal-Dölbau – both in Germany), Alho four factories (Morsbach, Coswig in Germany; Wikon in Switzerland, Mitry Mory in France), and Cadolto two factories (Cadolzburg and Krölpa in Germany). In the following steel-based prefabrication will exemplarily be outlined by the example of Cadolto's manufacturing process.

Market Share/Business Strategy

Codolto supplies prefabricated buildings with a prefabrication degree of up to 90%. In contrast to the companies and systems mentioned in Section 4.1.1 and that prefabricate family homes (wood- or brick-based), Cadolto's modules (based on three-dimensional steel frame units) are used for functional, nonresidential buildings such as office and administration, education (kindergartens, schools), research (laboratories) buildings; and buildings related to healthcare (homes for the elderly, hospitals), services, or retail (bank facilities, shops, mobile offices); and buildings related to the hotel and restaurant industry (hotels, cafeterias, canteens). A specialty of Cadolto

Figure 4.1. Prefabricated modules of Cadolto are fully pre-equipped with all utilities (electricity, aeration, etc.). (Photo: Cadolto Fertiggebäude GmbH & Co. KG)

is the super-fast prefabrication of hospitals, and magnetic resonance imaging (MRI) and operation rooms.

In particular, Cadolto has close relationships with suppliers of medical technology, laboratory equipment, and equipment for homes for the elderly as it integrates these items into rooms or building parts completely finished in the factory. Suppliers honour the achieved quality and precision and therefore the integration of the equipment is made possible within the factory of Cadolto. Complex medical technology (e.g., magnetic resonance tomography [MRT]) or laboratory equipment in particular demands that individual parts of that equipment are set up with maximum precision. The finishing of rooms equipped with such technology also has the advantage that the need for highly specialized workers to position, set up, and put this equipment into operation on-site is reduced. This is a particular advantage for the companies, suppliers, and developers considering the fact that more and more complex equipment is being demanded in countries such as Russia, Hungary, and China, where highly specialized workers able to perform a precise and reliable setup are limited. For this reason, Cadolto has sent an increasing number of fully equipped operation or MRI rooms to Russia and China.

A further advantage is that the described type of prefabrication allows for hybrid buildings that are made up of parts built in a conventional way by a local construction company and modules coming from the Cadolto factories. Thus, it becomes possible to equip conventionally built buildings (e.g., hospitals) with prefabricated modules containing high-tech equipment (e.g., operating or MRI rooms).

The Manufacturing Process

Cadolto is prefabricating steel frame units of various types and sizes (Figures 4.1 to 4.3, Tables 4.2 and 4.3). The units can be completed and finished by them in their factories (including standardized or individual facades and interior elements). Alternatively, this can be done on the basis of steel frame units by other specialized companies or carpenters. Similarly, planning can be done either completely or only partially by them. In many cases the exterior design is coordinated by the customer's architects or planners and detail planning and individual fabrication is then done by Cadolto.

Cadolto (as well as Kleusberg and Alho) also offers downloadable software for planners and architects for configuration of the building. With this software, planners arrange buildings on the basis of the available steel unit sizes. The built-in equipment and finishing, along with other infill, can be configured individually and then be finished by Cadolto itself, or by another company.

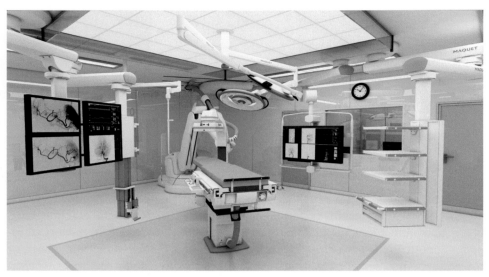

Figure 4.2. A speciality is the prefabrication and shipment of prefabricated hospitals. Cadolto preinstalls, for example, complex (e.g., MRI) equipment for the operating rooms already in the factory. (Photo: Cadolto Fertiggebäude GmbH & Co. KG)

Lead Times

The manufacturing of bare steel frame units takes about one week and the finishing of the modules, including the integration of equipment, between two and four weeks. On the construction site, buildings (e.g., small hospitals) can be assembled within three to four months. However, between factory completion and on-site assembly, modules often stand still (in manufacturing this is considered as idle time) for a month or more.

Product

The measurements of the bare steel frames are restricted primarily by transportation demands. Cadolto, however, works closely with specialized logistics providers and can thus ship large units (i.e., more than 30 metres in length, weight up to 50 tons). The steel frame is a stiff and bearing frame that can be equipped with interior infill (insulation, cables, pipes, electricity, equipment, furniture) as well as with exterior finishing. A modular design–oriented (steel frame unit) structure allows a multitude of floor layouts including the generation of elevator and staircases elements.

Off-Site Manufacturing

The off-site manufacturing process is mechanized and labour-based. However, despite the lack of automation, the process is highly systemized with a flow line–like organization. Employees work in a structured factory environment and are assisted by working platforms, ceiling guided welding devices, balancers, overhead cranes, and mobile task-specific workbenches.

At Cadolto suppliers provide plasterboard in the correct amount and dimensions, avoiding nonreusable cutoff waste completely. Furthermore, Cadolto employs all necessary craftsmen to complete a building. In the finishing area, craftsmen are

Figure 4.3. View of Cadolto's factory. (Photo: Cadolto Fertiggebäude GmbH & Co. KG)

Table 4.2. *Off-site manufacturing process: production of modules in the factory*

a) Generation of the steel frame unit

b) Interior finishing

c) Exterior finishing

Photos: Cadolto Fertiggebäude GmbH & Co. KG.

Table 4.3. *Logistics, on-site assembly, and last finishes*

a) Delivery of the prefabricated modules from the factory to the site

b) On-site installation of the prefabricated modules

Photos: Cadolto Fertiggebäude GmbH & Co. KG.

assisted by task-specific mobile mini-workbenches. Material supply for the finishing area is organized centrally from a supply shop that also equips the mini-workbenches with tools and materials.

On-Site Assembly
On the construction site, the crane (able to lift in some cases as, e.g., operating rooms) lifts units of up to 50 tons with up to five workers needed to position and

Table 4.4. *Kleusberg GmbH & Co. KG*

Business volume/year	€110 million
Output/year	—
Staff	450 employees
Construction and material	Multipurpose steel units
Range of products	Hospital, laboratory, office building, homes for the elderly, temporary and middle-term container solutions; room–in–room systems, exhibition systems
Distribution area	Germany, the Netherlands, Belgium, France, Switzerland, Austria, Poland
Price range	—

assemble the modules. Units are delivered to the site on the basis of fixed time schedules just in time (JIT) and just in sequence (JIS) in 30-minute cycles. The process of unit pickup from the truck and positioning takes, depending on unit size, approximately 30 minutes per module. All four corners of each unit have factory inbuilt steel connector elements allowing easy and fast pick up from the truck as well as connection to other units on the site. For adjustment of the units after positioning, human workers use small hydraulic presses.

Analysis of Selected Companies

The most representative German companies active in steel prefabrication were chosen for comparison. Kleusberg (Table 4.4), Alho (Table 4.5), and Cadolto (Table 4.6) have developed a broader experience within this field.

4.2 United Kingdom

4.2.1 History

In the United Kingdom, the use of LSP methods has been developed and implemented in many various forms for many years. From the construction of the Crystal Palace for the Great Exhibition in 1851, to the houses for the heroes movement in the 1940s and the introduction of council housing in the 1950s (Hollow, 2011), the prefabricated construction sector in the United Kingdom has had its ups and downs

Table 4.5. *Alho Systembau GmbH*

Business volume/year	€95 million
Output/year	—
Staff	300 employees
Construction and material	Multipurpose steel units
Range of products	Offices, hospitals, homes for the elderly, hotels, kindergartens, schools, universities, medium-term solutions
Distribution area	Germany, Belgium, Switzerland, France, Luxembourg, the Netherlands
Price range	—

Table 4.6. *Cadolto GmbH & Co. KG*

Business volume/year	More than €100 million
Output/year	About 1200 large modules per year
Staff	About 300 employees
Construction and material	Multipurpose steel units
Range of products	Hospital, laboratory, office building,
Distribution area	Germany (50%), foreign countries (50%, i.e., Belgium, Switzerland, Russia)
Price range	—

through World War II, housing boom, and economic downturn (Short, 1982). Throughout these different eras, the building prefabrication sector has successfully overcome many issues such as lack of skilled labour forces and short supply of building materials. However, with the ever-growing popularity of industrialized construction in the 1960s, the Ronan Point incident (collapse of a prefabricated corner part of a high-rise building) displayed the controversy within the building prefabrication industry (Pearson and Delatte, 2005). Furthermore, the incident had a negative impact on public confidence in prefabricated construction as a whole.

4.2.2 General Overview

In the United Kingdom, the prefabrication sector is focussed primarily on public and commercial sectors such as healthcare, education, military, offices, hotels, and supermarkets. Recently, because of an increasing demand of affordable and social housing, the government has shifted its priority to the development of prefabricated construction (Loison-Leruste and Quilgars, 2009). In general, the UK building prefabrication industry can be divided into four main market categories: timber frame,

Table 4.7. *Main British prefabrication companies*

	Since	Business volume/ year[a]	Output/ year	Employees	Range of products	Price range[b]	No. of factories
Terrapin	1949	28		200	Portable modular building and timber frame building components	–	1
Caledonian Modular	1964	12		300	Steel frame modular building system, bathroom pod	–	7
Kingspan	2001			300	SIP and timber frame building system "TEK"	–	1
Yorkon	1981	16		800	Steel frame modular building system	–	1
Laing O'Rourke	1978	4000 overall		17,352 overall	Modular home, precast concrete building system	–	1

[a] In million GBP.
[b] In GBP per square metre.

light gauge steel framing systems, precast concrete systems, and structural insulated panels. In 2011, the prefabrication market was estimated to account for around 2% of the total construction market (Taylor, 2011).

4.2.3 Companies

There are approximately 40 building prefabrication companies in the United Kingdom. The collective yearly output from 2011 was 6000 units (see, for example, SDI, 2015). Most of them are relatively small companies with little effective coordination or partnering. The major players in the building prefabrication industry are Laing O'Rourke, Yorkon, Caledonian Modular, Terrapin, and Kingspan off-site. In contrast to the building prefabrication industry in Germany or Japan, United Kingdom companies are supplying panelized building components and three-dimensional modules predominantly as sub-components for conventional construction projects rather than completely prefabricated buildings. Table 4.7 outlines the performance of selected major players in the building prefabrication industry in the United Kingdom.

Laing O'Rourke (Figure 4.4) is the largest privately owned construction firm in the United Kingdom. It produces a range of products including precast concrete walls, floors, columns, and beams for their projects throughout the United Kingdom. Recently, the company has been aiming to move into the house building sector, based on its expertise in using off-site technology to provide a range of sustainable modular housing solutions. Highly automated, production line–based solutions were deployed recently at Laing O'Rourke's precast concrete factory (Explore Manufacturing Plant, see **Volume 1, Section 4.3.3** for further information).

Figure 4.4. Laing O'Rourke's production facility. (Photo: Laing O'Rourke)

The Yorkon building system (Figures 4.5 to 4.7) is manufactured in York at a large 250,000 square metre production facility that belongs to the Portakabin Group, the parent company of Yorkon. Yorkon provides a full range of modular construction services, from planning, design, production, all the way to completion of the project. As part of the Portakabin Group, Yorkon is one of the United Kingdom's leading modular building manufacturers; the company also offers a range of highly

Figure 4.5. Yorkon: production of panels. (Photo: Yorkon Ltd.)

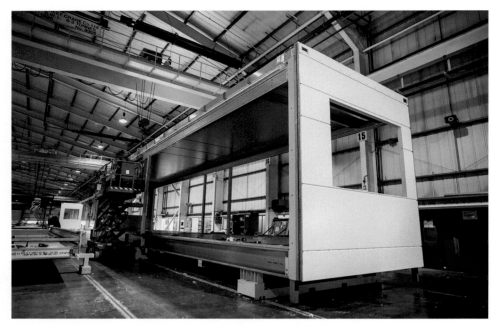

Figure 4.6. Yorkon: production of units. (Photo: Yorkon Ltd.)

innovative solutions for steel frame modular buildings such as schools, hospitals, and supermarkets.

Kingspan off-site is one of the subdivisions of the Kingspan Group. The company launched TEK Building System in 2001, after taking over German-based TEK Haus. The TEK building system contains high-performance structural insulated panels

Figure 4.7. Yorkon: Cambourne College. (Photo: Yorkon Ltd.)

connected by a dedicated joining system for walls, roofs, and intermediate floors. The system features high design flexibility, excellent U-values, and a sustainable approach in housing construction. TEK is still one of the most popular products provided by the company. Kingspan TEK product offerings were recently incorporated into Kingspan Insulation Limited.

Caledonian Building Systems Limited, based in Carlton-on-Trent, Nottingham-shire, provides preengineered buildings and off-site construction services. The company specializes in the design, manufacturing, and construction of permanent or temporary multistorey buildings using steel frame modular structures and volumetric building modules/units. Their products serve health, education, custodial, residential, hotel, and commercial office sectors. The company also produces prefabricated bathroom modules as components/modules for hotels and student accommodations.

Terrapin is one of the pioneers in the off-site construction sector in the United Kingdom. In 1949, Terrapin developed the prefabricated bungalows that solved the post-war housing shortage problems. In the early 1990s, the company launched a modular, prefabricated system with a rigid light steel frame for residential and key worker accommodation. The company currently offers a full range of off-site products: volumetric, panelized, and steel-framed modular systems, as well as full design and project management service across the country.

4.2.4 Manufacturing Methods

Most of the UK-based building prefabrication companies use highly auto-mated computer-based manufacturing facilities to ensure high-quality produc-tion. Computer-aided design/computer-aided manufacturing (CAD/CAM) are com-monly used throughout the industry. Huge technical investments have been poured into the industry, aiming to improve building envelope advancements, enhance pro-duction processes and communication technology in manufacturing, and develop products to comply with product certifications across Europe.

4.2.5 Conclusion

The housing prefabrication industry in the United Kingdom has grown slowly because of many circumstances. The scope for application of prefabricated building system is still limited. Many prefabrication manufacturers are running below produc-tion capacity because of lower demand. This trend might be caused by the popularity of second hand housing market. Overall, there are 8.8 million (38%) dwellings in England that were built before 1945; more than half of these (4.8 million) were built before 1919, according to an English housing survey in 2011 (DCLG, 2013). On the other hand, the residential buildings market still considers prefabricated houses as more expensive than traditional on-site methods of construction. In addition, most prefabricated solutions are considered as "nonstandard construction", and because the regulations of many financial institutes do not easily permit mortgages on build-ings built with a nonstandard construction methods, this has a negative effect on the prefabrication industry. However, with increasing demands for social housing and growing government interests, the UK prefabrication market faces a rather optimistic future.

Table 4.8. *Major Spanish prefabrication companies*

	Since	Business volume/ year[a]	Output/ year	Employees	Range of products	Price range[b]	No. of factories
Compact habit	2009	12			Residential, health		1
Modultec	2001	30	Capacity for 200,000 m^2	110	Any		1
Egoin	1989	10	15,000 m^2	54	Residential, educational		1

[a] In million euro.
[b] In euros per square metre.

4.3 Spain

4.3.1 History

In 1996, the government liberalized the market in the construction field. The European Union also provided funding for infrastructure. Spain's tourism industry requires many services such as hotels, apartments, and other various infrastructures. There has, however, been little investment on R&D in the country, especially in the construction sector. Construction processes remained oriented towards old techniques. Cheap workforce was used and many foreign workers were utilized as labour force.

4.3.2 General Overview

Table 4.8 gives an overview of the most important Spanish prefabrication companies.

4.3.3 Companies

Although there are many big construction companies located in Spain (ACS, FCC, OHL) that operate all over the world, they often do not make a significant effort in renewing prefabrication techniques. There are few prefabricated module/units producers and builders. The ones chosen here are the ones that have designed, produced, and sold their product on a regular basis. Compact Habit specializes in concrete components, Modultec works mainly with steel, and Egoin works with wood- based components/modules/units.

4.3.4 Manufacturing Methods

In the analysed cases, the manufacturing method is based on a flow-line and production-line–focused manufacturing layouts for producing a the components or modules.

4.3.5 Conclusion

Although companies lack a little bit behind with the deployment of prefabrication, research and development in building component manufacturing as well as in modular construction and prefabrication is conducted by many universities as well as by the research centres of large construction firms.

4.4 China

4.4.1 History

Building prefabrication began in China in 1949 and developed radically during the 1950s (Ho, 2005). As the aim at that time was to catch up with Western developed countries, the state government adopted prefabrication technology from the former Soviet Union. Large panel systems were implemented nationwide; in the following decade this building method became the solution for urban housing development in China. However, the structural and workmanship deficiencies in the system were ignored. The consequences of the Tang Shan earthquake in 1976 were catastrophic; many buildings were destroyed because of structural failure. Housing demands increased dramatically after economic and building market reform in the 1980s, and various building types and on-site and off-site prefabrication technologies were introduced. Over the next two decades, building prefabrication development slowed despite the government's "Housing Industrialization" strategy. The industry was mainly focussed on acquisitions from the booming real estate market rather than improvement of productivity, technology, innovation, and sustainable development. Recently, owing to the rapid urbanization and shortage of skilled construction labour, the Chinese construction sector is once again favouring the development of LSP.

4.4.2 General Overview

In China, the building prefabrication sector has been focused primarily on producing building elements and components such as precast shear walls, floors, and prestressed concrete columns and beams. Recently, the Chinese Ministry of Construction realized the necessity of prioritizing the design of industrialized building, production of standardized building elements, and off-site construction methods. Therefore, new policies and targets (PDO, 2005) have been published to promote prefabricated construction methods, especially precast concrete and steel frame structure, and improvements in prefabrication quality and assembly technique. Many pilot projects and pilot companies were encouraged to experiment with building prefabrication methods in an unprecedented scale that the world has never seen.

4.4.3 Companies

There are numerous building prefabrication companies in China. In general they can be divided into four categories: state-owned enterprises, private companies, urban and rural collectives, and real estate companies. Uniquely, in China, many real estate

Table 4.9. *Main Chinese prefabrication companies*

	Since	Business volume/ year[a]	Output/ year	Employees	Range of products	Price range[b]	No. of factories
One Star Group	1998	+200 million	+1.5 million m^2	1000	Precast concrete, wall, floor panel and steel building structure	Approx. 500	2
Broad Group	1988	+1000 billion	+10 million m^2	4000	Steel frame building system	Approx. 1000	1
Vanke	1999	+100 billion	+120 million m^2	35,000	Precast concrete, wall, floor panel	Approx. 500	Real estate developer

[a] In RMB.
[b] In euros per square metre.

companies are directly involved in the design, production, construction, and R&D phases and have become major players in the field. Table 4.9 outlines the main Chinese companies that are active in the field of prefabrication.

One Star Group

Founded in 1998, One Star Group is one of the biggest manufacturers of steel frame building structures, precast concrete shear walls, precast ribbed floor panels, and precast building components in Shandong Province. The company developed a prefabricated building system consisting of both steel frame structures and precast concrete shear walls for the building envelope, and interior finishing and services that are self-contained as an infill layer. Recently, the company has been identified by the state government as one of the pilot companies whose aim is to improve the construction industrialization process in the region. The production facility has more than 2 million square metres annual capacity.

Vanke

Vanke launched its building prefabrication production in 1999 and is now China's biggest real estate developer as well as largest precast concrete building element producer, with a turnover of 100 billion RMB (€10 billion) in 2010. The amount of floor space constructed was more than 12 million square metres. The company produces precast elements for residential and commercial projects across China as well as in Hong Kong, the United States, and Singapore. It has a production output of more than 1 billion square metres annually. Vanke is one of the leading companies in the production and innovation of precast concrete shear walls and prestressed building components in China.

Vanke is known as one of the market leaders in designing and producing prefabricated building components in China. They have developed a range of precast concrete building elements and sustainable concepts proposed to cope with the implementation of various ranges of industrialised building systems and rapid construction demands. The company was commissioned by the state to erect a number of pilot projects across the country. Some of their key products are listed below:

- Precast walls, floor elements
- Prestressed beams and columns
- Bathroom modules, kitchen modules
- Precast balconies and stairwells

The company has recently started to develop the concept of "mini-residential units". The concept is focussed primarily on young buyers, and is to create compact but comfortable, fashionable, and convenient living environments.

Broad

Broad group is a private company, founded in 1988, and is the biggest manufacturer of central air conditioning and sustainable green products, including sustainable building solutions. Under the sustainable building division, the company has developed completely off-site manufacturable light steel framed building system, with a prefabrication degree of up to 93% (amount of work that can be completed in the factory; Broad, 2015). The system still holds the record of erecting a 30-storey hotel in just 15 days. Recently, the company proposed to construct a 220-storey building (Sky City) in less than 90 days. The company plans also to provide its steel frame system to the South American residential social housing market.

The steel frame system consists of several main parts: loadbearing structure, structural board and secondary structure, wall, celling, interior dividing walls, and services. The main structural board is 15.6 metres long (alternatively 11.7 metres or 7.8 metres), 3.9 metres wide and 0.45 metre thick. The design dimensions allow the components to be transported by lorry. In general, each lorry is capable of delivering two or three main structural boards. All wall and ceiling components, appliances, air-conditioning units, windows, doors, and internal supporting structure are placed on top of the main structural board. When the lorry arrives at the construction site, the entire load is then delivered to the prearranged assembly location and the workforce can carry out the on-site assembly tasks efficiently.

4.4.4 Manufacturing Method

In China, because of the scale of the country's development, prefabrication manufacturing methods are based primarily on precast and prestress techniques. Prefabricated steel structures are mainly used on high-end commercial projects. Because of the rising cost of steel, it is generally not utilized on a bigger scale. Despite these efforts, many prefabrication companies in China have upgraded and automated their production facilities in the past 10 years. However, compared to the West and also to Japan, the overall productivity and automation level in China is still relatively low. This is caused by the imbalanced development between different regions and state government involvement in the construction sector (Luo, 2004). Bigger enterprises are normally equipped adequately, and automation levels are higher than in smaller firms.

4.4.5 Conclusion

China first experienced exceptional economic achievements, followed by urbanization and increasing demand for housing. Recently, the state government has focussed on reforming the construction industry and made many efforts and investments in the hope that prefabrication and achieving construction industrialization would improve productivity, construction speed, quality, and building a sustainable future for the nation.

5 Large-Scale Building System Manufacturing in Japan

Japan has the most successful housing prefabrication industry in the world and has maintained this position for about 40 years. Today, the Japanese prefabrication industry manufactures about 150,000 entirely prefabricated housing units per annum with a continuously increasing degree of quality and embedded technologies. During peak times, the percentage of completely prefabricated houses in Japan was about 18 to 19%; the total market share of prefabricated elements is much higher and probably the highest one in the world. Besides the large and consistent market share, it is remarkable that the industry supplies the upper market segments (rather than the lower market segments) with fully customized, earthquake-resistant high-tech buildings.

In Japan, buildings constructed from prefabricated elements are not significantly more expensive than conventionally built buildings and include in most cases proactive building maintenance services bundled to them (i.e., service plans that not only repair damages when they occur but also consistently inspect and maintain the building) to guarantee that the building and its subsystems are consistently functioning and not causing any inconveniences to inhabitants or owners. To be able to provide outstanding quality, almost all manufacturers use automated machines and robot systems in their factories and organize their means of production along a production chain or even production line. The average salaries paid by Japanese prefabrication companies are among the highest in the Japanese general industry. Most Japanese prefabrication companies have no strong roots in the construction industry but rather originate from multinational chemical, electronics, or automotive companies.

Currently, the Japanese LSP industry advances directed towards adding and emphasizing complex additional functions and services playing a major role in the country's disaster prevention and disaster management strategy and developing and delivering entire "smart" cities that are sustainable, affordable, and assistive. Japan's prefabrication industry changes the notion of buildings recognized as simple "construction" products towards the notion of buildings recognized as complex high-tech products with completely new, service-oriented value creation possibilities – and its advanced manufacturing capability is the backbone for this evolution.

To be able to fully outline and explain the success of the Japanese prefabrication industry, its key elements, and machine-based manufacturing technology, first the background, development, and strategy of the industry are discussed (Section 5.1) before, second, identifying aspects of robot-oriented design (ROD) and management deployed in the industry (Section 5.2). Then, the focus is on the analysis of the typologies of manufacturing processes and layouts used by companies (Section 5.3). Next, a detailed outline of the 13 most important companies and their products and manufacturing systems is presented (Section 5.4). The chapter concludes with a discussion of current and future developments and strategic essentials in the Japanese LSP industry (Section 5.5).

5.1 Background, Development, and Strategy of the Industry

Although, as shown later, the manufacturing technology of Japanese prefabrication companies is key, their success is based on multiple factors and the exploitation of complementarities. To understand the circumstances in which these advanced and automated manufacturing processes are embedded, first, in this section, an overview of companies, turnover, output, employees, prices, salaries, and factory distribution is presented (Section 5.1.1). Then, the diffusion of karakuri technology into the prefabrication industry, the influences of local and cultural specifics on the evolution of standardization, the drivers for prefabrication in Japan, and the first approaches to mass production are discussed (Sections 5.1.2 to 5.1.7). The exchange of knowledge within the industry and the evolution from Sekisui Heim's M1 to the deployment of cutting edge know from general manufacturing industry in the form of Toyota Production System (TPS) principles and enterprise resource planning (ERP) systems in the prefabrication industry are also outlined (Sections 5.1.8 to 5.1.10). A sub-section giving a comprehensive timeline of the evolution of prefabrication in Japan concludes the section (Section 5.1.11).

5.1.1 Overview Companies (Turnover, Output, Employees, Prices, Factories)

In particular, the major prefabrication companies run several factories distributed over Japan. Each factory produces predominantly the types of houses required in the surrounding region (e.g., a factory in Hokkaido is optimized for building types demanded in the region and for buildings with additional layers of insulation). A high concentration of factories can be identified in the Nagoya–Osaka area. This is strategically advantageous, as the area is situated right in the centre of Japan and factories located there can supply both the Honshu and Shikoku areas. Major companies such as Sekisui House, Daiwa House, and Sekisui Heim operate additional factories in Kyushu and in Hokkaido.

Table 5.1 shows that major Japanese prefabrication companies are (e.g., compared to German prefabrication companies) relatively young, with (in the case of Toyota, Sekisui Heim, and Sekisui House) absolutely no background in the construction or housing industry. However, maybe because of this, they have shown a great willingness to change the product structure to fit the purpose of chain or line production, automation, and the buildup of an original equipment manufacturer (OEM)-like industry structure. The turnover of the listed housing manufacturing companies is in the billions and can be compared to the turnover of major German

Table 5.1. *Performance of the Japanese LSP industry in 2011*

	Since	Business volume/ year[a]	Output (houses)/ year[b]	Employees[c]	Range of products	Price range[d]	Factories
Sekisui House	1960	(1,530,577); 15,064	48,071	15,302	Steel Wooden	55–85 50–80	5
Daiwa House	1955	(1,116,665); 10,987	(43,000); 41,004	13,592	Steel Wooden	55–85 55–80	10
Pana Home	1963	(250,777); 2.467	10,753	4,264	Steel	50–80	2
Sanyo Homes	1969	(36,660); 361	1,010	668	Steel	55–65	1
Asahi Kasei – Hebel House	1972	(452,000) 4,447	16,231	5366	Steel	70–80	2
Misawa Homes – Assembly	1967	(378,500) 3,724	12,353	(8,917) 694	Steel Wooden	60–80 45–90	12 (Hybrid: 1)
Mitsui	1974	(216,838) 2,133	5,230	2,326	Wooden	50–80	7
Tama Home	1998	(153,719) 1,512	9,216	2,784	Wooden	30–50	1
Muji	1989	No numbers revealed	No numbers revealed	50	Wooden	55–80	Subcontractors
Sekisui Heim	1947	(449,000) 4,417	(14,600) 10,200	8,820	Steel Wooden	60–85 60–80	6
Toyota Home	1975	(131,871) 1,297	(5,400) 4,142	3,402	Steel	45–80	3

[a] In million euro (in million yen).
[b] (. . .) = including subcompanies, subcontractors, and related companies and brands.
[c] Not including, e.g., on-site assembly subcontractors.
[d] In 10,000 yen per Tatami unit.

Sources: Yearly financial reports of the mentioned companies. Sekisui House: www.sekisuihouse.co .jp; Daiwa House: www.daiwahouse.co.jp; Sanyo Homes: www.sanyohomes.co.jp; Asahi Kasei – Hebel House: www.asahi-kasei.co.jp; Misawa Homes: www.misawa.co.jp; Mitsui: www.mitsui.com/jp; Tama Home: www.tamahome.jp; Muji: http://ryohin-keikaku.jp; Sekisui Heim: www.sekisui.co.jp; Toyota Home: www.toyotahome.co.jp; Pana Home: http://www.panahome.jp; all last accessed October, 2013. Data from 2011.

and European contractors. The table further shows that the rate of work productivity (output/employee) in general becomes more advantageous with rising output numbers (Sekisui House and Daiwa House: 3.2 houses per employee per year; Sekisui Heim, Toyota Home, Pana Home, only 1 to 2 houses per year per employee). However, the work productivity of Sekisui Heim, Toyota Home, and Pana Home compared to Sekisui House and Daiwa House is not as poor as a first glance at the numbers might suggest. The official number of employees doesn't include a large number of employees of local subcontractors who support the on-site assembly of the steel- and wood-based panels. The work productivity of Sekisui Heim and Toyota Home more accurately reflects the reality, as up to 85% of the work is done in the factory, with fewer or nearly no subcontractors (which would not be shown in the employee's statistics) required. Given that price range and quality of Tama Home basically reflect the level of low-cost housing, the table shows that most companies operate mainly in the high-price and high-quality market segment. Currently, Sekisui House is considered to be the company that delivers a high percentage of very expensive buildings with the highest quality. This is also reflected by the fact that although "only" 10,200 buildings were manufactured in 2011 by the core company itself, the turnover nevertheless accounted for 4417 thousand million euros this year.

Further, Japanese salary statistics show that the average salary that Japanese prefabrication companies pay is among the highest in the construction industry and exceeds the salary paid by conventional construction companies as well as by most major contractors.

The following list compares average annual salaries per employee across two companies and three career positions (according to Nikkei BP, 2011):

1. Sekisui House (2011): €70,416.79 after taxes
2. Daiwa House (2011): €75,345.51 after taxes
3. Civil/construction engineer in Germany: €47,500.00 after taxes
4. Junior managing director in the German construction industry: €70,831 after taxes
5. Junior electrician in the German construction industry: €23,357 after taxes

Major Japanese prefabrication companies are members of the Prefab Club (joined by Sekisui Heim, Toyota Home, Misawa Home, Daiwa House, Sekisui House, Asahi Kasei, Prefab Club, 2013). The Prefab Club has a kind of exclusive relationship with the Japanese government concerning the supply of (temporary and short-term) housing for disaster areas. For example, after the March 11 Tohoku earthquake, the prefabrication industry, through the Prefab Club, delivered most of the required housing (about 40,000) – non–Prefab Club members and conventional house builders faced a disadvantage. The delivery of houses to disaster areas (Japan is frequently hit by disasters) increases the companies' output and business volume and allows them to better utilize their high-capacity factories.

5.1.2 Japan's Prefabrication Industry Today and Tomorrow

Japanese prefab makers altogether build about 150,000 housing units per annum (see also Linner & Bock, 2012a on which Section 5.1.2 is partly based). For comparison, in Germany in economically weak years the same number of building permits are issued for private housing in conventional and prefabricated construction combined. Already in the 1970s, Sekisui Heim's legendary M1 reached a sensational annual production of more than 3000 buildings. High annual production on a steady level – along with the interest of the companies behind, for example, Sekisui Heim and Toyota Home to explore new markets – allowed for the investment in the advancement of component systems, expensive manufacturing technology (e.g., production lines, automation and robotics, advanced logistic systems), and extended customer services, which characterize the uniqueness and strength of the Japanese prefabrication industry today. Based on this, the productivity of this industry has become so high that – depending on the degree of capacity utilization the economic situation allows in a specific year – three to four customized buildings per employee can be realized annually.

Table 5.2. *Production peaks of the main players in the prefabrication industry*

	Sekisui house	Daiwa house	Sekisui heim	Toyota home
Units per year (peaks)	78,275 (1994)	44,500 (2007)	34,560 (1997)	5024 (2006)

Source: Yearly financial reports of the mentioned companies. Sekisui House: www.sekisuihouse.co.jp; Daiwa House: www.daiwahouse.co.jp; Sekisui Heim: www.sekisui.co.jp; Toyota Home: www.toyotahome .co.jp; all last accessed October, 2013.

Japan's prefabrication industry is among the strongest worldwide. However, it has undergone a steady change and decline since the 1990s. A maximum production peak was reached in the mid-1990s, with approximately 600,000 newly constructed housing units. Later in 2000, about 450,000 units were constructed, and in 2009, construction declined to about 300,000 units. During the peak times, the percentage of prefabricated houses (in % of all built houses) was about 18 to 19%, which decreased to just 13 to 15% at present, depending on the region. However, a high amount of prefabricated elements are being used in conventional construction, which increases the actual percentage of prefabrication in the whole building industry, although it is hard to express this phenomenon in numbers. The prefabrication of entire buildings could be broken down into about 80% steel-based building kits, 15% wood-based building kits, and 5% concrete-based building kits.

Sekisui House, which is still the one of the main players in Japan's prefabrication industry today, reached its peak in 1994 with production of 78,275 housing units (Table 5.2). At that time, Sekisui's share of the total building construction market in Japan was 5.3%. It is interesting that both Sekisui House and Daiwa House, the second largest player in Japan's prefabrication industry, today try more and more to address the decline in the market by transitioning into a developer position. Houses and apartments are developed, planned, and constructed for later rental to customers. These houses and apartments are also based on the companies' prefabricated housing kits and ensure that the capacities of expensive automated manufacturing facilities are utilized to a maximum.

To manage the decrease in demand and to build up new ways of value creation, all of the main players are focussing now also more on the buildings' utilization phase, building performance, and building life-cycle services. Sekisui House is working on modular upgrade packages for older model lines, allowing their owners to upgrade easily and continuously, both building design and performance. Daiwa House cooperates with companies like Cyberdyne (HAL) and Toto (Intelligent Health Toilet) to develop assistance technologies and advanced healthcare services embedded in or related to their homes that can be sold with them. Sekisui Heim is trying to gain a leading position in matters of sustainable low-energy houses and has been building up a system for reverse logistics and building recustomization with its Reuse House System. Furthermore, it has built up, together with its suppliers, a BIM-based information management system that allows dynamic data management of the building, components, customers, maintenance, and services. Toyota Home is gradually improving its graded warranty and service models with the intention of providing long-term building maintenance and facility management. If Japan's prefabrication industry continues towards a successful implementation of the aforementioned life-cycle services and connects them to approved mass customization structures and customer integration strategies, a new prototype of the construction industry based on the fusion of mass customization and building/household–related services might be the result. The focus won't then be on material, resource, and labour input exclusively any more but rather on long-term customer relations and product service systems.

Moreover, the high productivity rate of large-scale industrialization and automation practiced in Japan must be mentioned. The main players in Japan's prefabrication industry managed to build about three to four houses per employee per year. Additionally, in contrast to the prefabrication industry in Europe and the United States, Japan's prefabrication industry has dedicated itself to a premium class strategy in general. An analysis conducted by the authors of this chapter in 2007 revealed that the majority of Japanese prefabricated homes have square metre costs significantly above the average. Research and interviews with sales professionals have shown that most of the customers of prefabricated homes have a high income and high social status. For those people, quality-oriented and earthquake-resistant homes, fabricated to individual needs by the speed and precision of sophisticated automation and robotic systems, accompanied by advanced warranty and maintenance services, are much more affordable than for those with an average income. Prefabricated homes in Japan can be compared with premium class cars and status symbols such as BMW, Mercedes, and Audi. Furthermore, the manufacturers invest extensively in their research and development centres as well as in the implementation of new technologies and services. Accordingly, today's Japanese prefabricated homes are value-added and quality-oriented products accompanied by a multitude of advanced services.

5.1.3 Karakuri Technology Diffusion in Japan

Karakuri mechanisms are accounted as predecessors of modern automation and robot technology (Wißnet, 2007). Karakuri ningyō are mechanisms that are actuated by ropes, springs, and gear systems to manipulate things such as hands, objects to carry, bows for shooting small arrows, and others (Figure 5.1). Manipulation can take place automatically (by springs running the mechanism) or by direct human manipulation via ropes. In contrast to the automata developed in Europe, karakuri is a widespread handcraft in Japan and karakuri mechanisms can be found in many ordinary homes as art and entertainment devices.

One of the most famous karakuri mechanisms is the chahakobi karakuri puppet actuated by springs and gears. The puppet is able to serve tea to a guest. The mechanism is initiated once a teacup is placed on the serving plate and interrupted once the guest picks up the cup. Additional puppets manipulated by ropes and gears were widespread. Although developed much earlier, Japanese karakuri mechanisms (16th century) used a similar systemic way of combining machine parts such as later 18th/19th century Monge, Hachette, and Borgnis in Europe (see also **Volume 1, Section 3.4.1**).

Bock (2011) describes karakuri as an elementary technology that was handed down from generation to generation – not only the concept of manipulation, but also the tools that were used. Karakuri technology diffused, and was used later in adapted

a) Karakuri puppet able to serve
tea to guests

b) Structure and actuation system
of Karakuri puppet

Figure 5.1. Karakuri technology represents an early form of manipulation.

forms in theatres, buildings, yatai, factories, and later on in automated construction sites. Further, Nomura (2012) stressed that karrakuri also built a basis for Japan's advanced spinning and weaving technologies that were later developed. He also stressed that within Toyota, the karakuri principle was undoubtedly transferred from weaving technology into modern car manufacturing technology. Based on this assumption it can also be said that Toyota, with Toyota Home using the TPS along with automation and robot technology to produce buildings on the production line, has transferred the karakuri principle to the production of buildings (see also Table 5.3).

5.1.4 Influences of Local and Cultural Specifics and Disasters

Kunio Maekawa established a technical basis for modern Japanese prefabrication (Matsukuma et al., 2006). Nevertheless, a number of local characteristics provided an additional impact on the development of prefabrication in Japan (see also Linner & Bock, 2012a and Linner & Bock, 2013). Also important for the wide acceptance of prefabrication was the ability of this culture to transform itself. A number of historical Japanese cities experienced dramatic changes during their (often rapid) evolution, for example, by natural disasters such as fires or earthquakes. Another requirement for Japan was to also change the capital with the introduction of a new emperor: entire cities were quickly moved and relocated, requiring strong and effective systems of measurements and sophisticated prefabrication approaches to allow rapid and affordable transformation and relocation processes. Thus major governmental and religious buildings were prefabricated using standardized and

Table 5.3. *Stepwise transfer and diffusion of karakuri principles from karakuri ningyō to the Toyota production system*

Karakuri ningyo
Mechanisms that are actuated by ropes, springs, and gear systems to manipulate things such as hands, objects to carry, bows for shooting small arrows, and others. Manipulation could take place automatically (e.g., by springs running the mechanism) or by direct human manipulation (e.g., via ropes).

Early spinning and weaving technology
Similar to karakuri mechanisms early weaving technology proceeded to a complex manipulation of mechanisms by multiple ropes.

Loom technology and Jacquard mechanism
From 1800 on looms were equipped with Jacquard mechanisms, and steam engines supplied continuous and uninterrupted power. Looms thus became self-acting and automatic machine systems. Ropes-, belts-, and threads-based manipulation played a major role in those mechanisms.

Intelligent automation: type G automatic loom
Toyota's type G automatic loom (1924) was the first loom with a proprioceptive (mechanical) sensor system that could stop automatically in case malfunction during weaving occurred and thus eliminated the need for supervision. The invention was later licensed to the United Kingdom and the money received for the license built the basis for building up Toyota's automotive section.

3D automatic manipulation technology
Toyota also experimented with complex 3D loom technology that can be seen as a predecessor of later 3D manufacturing and manipulation technology at Toyota Motors and Toyota Home.

Computer-controlled high-speed looms
Toyota integrates high-speed air jet loom technology with computer–controlled, allowing basically any textile pattern to be manufactured within the looms. Modern looms are able to operate the shuttle thousands of times per minute.

Fully automatic flexible body line
Toyota was one of the first companies worldwide that used automatic and robotic manipulators extensively (since 1971). In 1985 Toyota introduced the flexible body line (FBL) equipped with a multitude of welding robots and thus introduced fully automatic car body assembly. Ropes-, belts-, and threads-based manipulation was substituted by other manipulating and actuating mechanisms which, however, fulfilled a similar role.

Toyota Home and Toyota Production System
In the 1980s, Toyota develop its housing business with Toyota Home and started to produce prefabricated houses, transferring the Toyota Production System (TPS) from its automotive section to the industrialized and production line–based manufacturing of buildings. During the following decades, all other main players in the prefabrication industry followed this newly set trend, installing the basic ideas of TPS in their plants, products, and organizations. With the TPS also Toyota's know how in machines, assembly and manipulating/actuating mechanisms was transferred into building production.

industrialized construction methods. In contrast to European cultures, where a transformation of historical buildings is generally connected with a loss of local spirits (genius loci), the Japanese culture has developed a spirit of continuous renewal. A major gash in Japans history occurred in 1867, when its stability was harshly tested. After centuries of international isolation, the country fostered its opening. This was followed by a rapid transformation into a modern industrial nation to maintain its power as an independent state. The next challenge occurred at the end of the Second World War, when almost 30% of Japan's housing had been destroyed. Owing to the dramatic housing shortage, immediate and rapid deployment of shelter was again required – mostly prefabricated and, at the beginning, characterized by poor quality. These uncomfortable living conditions dominated Japan in the postwar era. After the demand was covered, the people gradually started to increase their standards by asking for safe and durable high-quality houses. Again, one of the answers was prefabrication and the industry managed to transform itself from delivering prefabricated homes with poor quality to a premium class strategy of delivering individual, earthquake-resistant, and service-accompanied homes. After the reconstruction in the 1950s, Japan prospered in the 1960s and 1970s. During these two decades, the "nuclear" family became increasingly predominant, whereas the traditional form of living in a three-generation home began to disappear. The new trend became a new type of home for the salaried man and his family – a detached house with gateway, gable, and garden. This development was also politically supported by the United States, which was convinced that individual home ownership was a strong weapon against communism. The large number of imitators caused urban sprawl and available land became scarce and expensive. So, in the early 1980s, when Japan gained strong economic power, the price of land went up rapidly and the dream started to fade. Consequently, the traditional multigenerational home was considered as efficient, space saving, and cost-friendly and began to regain popularity. Often multigenerational homes of that period were designed as three-story buildings and each story was assigned to a generation, beginning on the ground with the oldest. To follow this change of demand, new types of prefabricated houses had been developed, and again prefabricated, modular structures and "infills "were introduced to the homes, allowing for flexibility and adaptability to meet the changing demands of the various generation living together over time. In addition to these social and economic aspects, the development of prefabrication has been influenced positively by a number of earthquake disasters, which again made it necessary to build large amounts of new houses rapidly. Factory-based prefabrication was also supported by the rapid development of the steel industry in the 1960s. As the major material used in the housing market, steel can provide certain advantages related to standardization and automation and robotics-based manufacturing (see also **Section 6.5.2** in **Volume 1**: Designing for Manufacturing with a Specific Building Material, and Section 5.2.3 in this volume). Sekisui, a company with roots in the chemical industry, was supposed to start up its housing business with plastic units – to enforce and extend its original business. External influencing factors induced Sekisui to reconsider those intentions. The development of the Japanese steel market as well as the oil crisis in the 1970s convinced the company to use steel, instead of plastic, as its major construction material for its modular units.

5.1.5 Roots in Chemicals, Electronics, and the Automotive Industry

The Japanese prefabrication industry has roots in traditional Japanese timber construction, the Japanese preference for standardization (Ken, Tatami), Japanese joining systems, and the Metabolic movement, and thus in the architecture and construction professions themselves. Natural disasters, relocation of capitals, big fires (e.g., Tokyo), and war periods/wars have, over hundreds of years, led to the

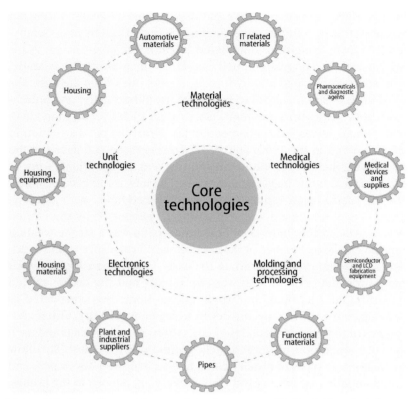

Figure 5.2. Overview products and product areas of Sekisui Chemical Corporation. Further information can be found on the corporate website (http://www.sekisuichemical.com/about/division/index.html).

refinement of building systems, standardization, and modularization so that buildings of all kinds could be optimized for frequent repair, relocation, or rebuild (see also Linner & Bock, 2013). However, the Japanese prefabrication industry also has strong roots in other, non-construction or building-related industries. It is symptomatic that most major successful prefabrication companies such as Sekisui House, Sekisui Heim (both originating from Sekisui Chemical Corporation), Pana Home (originating from Pansonic), and Toyota Home (originating from Toyota Motors Corporation) were set up by companies with a background in other industries. Of course these companies applied organizational forms and technologies (e.g., production line, automation, expert systems) they had already applied successfully in these non-construction industries. Sekisui Chemical Corporation, a chemicals company active in many industrial fields (see also Figure 5.2), set up and backed its housing business in the 1970s to create new markets for its core products (plastics, rubber), and Panasonic's Pana Home is, today, one of the biggest customers of Panasonic's home appliances, solar panels and smart home technology. Furthermore, the backbone, promoter, and beneficiary of large-scale prefabrication (LSP) in Japan was and continues to be the powerful Japanese steel industry (e.g., Kozai Club consisting of Kawasaki Steel, Kobe Steel, Nippon Steel, NKK, Nisshin Steel, Sumimoto Metal; Sakumoto, 1997).

5.1.6 Drivers for Prefabrication in Japan

The Japanese prefabrication industry is the only LSP industry worldwide that has reached a yearly output that allows investment in advanced manufacturing technology. The following drivers for the customer acceptance and need for prefabricated buildings in Japan can be identified:

1. *Affinity for technology in Japan*: Japanese people like robotics and high-tech products, and the fact that buildings are produced in a factory by automation and robotics enhances the perception of the reliability of the company enormously.

2. *Reliability of brands and big companies*: Japanese prefabrication companies are parts of huge (multi-business) companies with a long history and tradition (e.g., Toyota Home/Toyota, Sekisui Heim, Sekisui House/Sekisui Chemical). Customers can be sure that companies provide quality and that companies will exist in the next 50 years and thus be able to fulfil their service and warranty responsibilities.

3. *Marketing as a high-quality product*: Prefabrication companies see themselves as "premium brands". This makes buildings attractive for high- and low-income clients. Prefabrication companies apply "concept car"-strategy (as with R&D centres or concept houses such as the Toyota PAPI dream house).

4. *Continuous product improvement and innovation*: Prototypes of new or existing models are subject to experiments and testing in the companies' R&D centres (testing of behaviour towards rain, snow, fires, sound, earthquakes, typhoons, tsunamis, etc.). In contrast to conventional construction, prefabrication companies apply methods of professional product development (prototyping – testing – improving).

5. *Disaster/Earthquake resistance*: As all model series have been subject to a professional product development procedure including testing of behaviour of the building in case of an earthquake on shaking tables (platforms that allow simulation of earthquakes for 1:1 building prototypes), earthquake resistance – or resistance to other disasters – can to a certain degree be guaranteed. In addition, prefabrication companies offer the following technologies: vibration control panels (dampening systems integrated into structural elements) and base insulation systems.

6. *Customer integration*: Customers are invited into the companies' R&D centres for experiments and test living. On the basis of scientific data about users and their living habits, the company then generates the building's layout and functions. Further, the "ringi seido" culture also plays a role. This refers to nonhierarchical and informal decision making, bringing information from customers and production directly to management and product design, thus generating important feedback loops. Further, the marketing research divisions of Sekisui House, Daiwa House, Sekisui Heim, and Toyota Home investigate customer acceptance of the solution space in a six-month cycle. The results of these investigations are fed back into the design development stage and help to continuously improve the products. The building manufacturers gradually enhance the degree to which customers are involved in production and development issues.

7. *Advanced customer services and long-term relations*: Attracting and retaining clients is essential for the success of any customization-oriented strategy. Through

the customization process, companies receive detailed information about customers and establish strong relationships with them. Japanese prefabrication companies establish and maintain these relationships through

- Handover services
- Quality certificates and warranty
- Maintenance services of up to 60 years
- Upgrade services
- Renovation and rearrangement services
- Recustomization (e.g., System Reuse House of Sekisui Heim)

Figure 5.3 summarizes a study aimed at identifying and weighting the reasons that motivate clients to choose a prefabricated home. The study shows clearly that aspects such as the (perceived) reliability of large firms and brands (e.g., established large brands such as Sekisui or Toyota) as well as the superiority of quality and performance (which implies the factory automation aspect allowing companies to control quality as well as outstanding performance in terms of earthquake resistance and energy efficiency) can be considered most obviously as the most influential parameters. Noguchi (2013) argues that the outstanding performance of Japanese prefabricated homes is generated by the industry through intensive R&D and thus intensive reinvestment. Outstanding performance is the key marketing argument of Japanese prefabrication companies and clearly distinguishes prefabricated buildings from conventionally built buildings. This marketing strategy is reflected by the customers' income levels (Figure 5.4), showing that, in particular, among higher income classes prefabricated buildings are extremely popular.

5.1.7 First Approaches to Mass Production: Premos Home

The beginning of large-scale factory-based mass production of buildings began with Kunio Maekawa in the late 1930s and his PREMOS Home (Figure 5.5). He can be considered as the pioneer of modern Japanese prefabrication. Maekawa pointed out that a contemporary home should not just accommodate its inhabitants' resting, eating, and sleeping, but it should also facilitate cooking and cleaning, adequate light, ventilation, and heat. New American houses often included facilities such electric ranges, washing machines, refrigerators, flush toilets, heating, and insulation. Maekawa acknowledged that Japan was still a poor country and couldn't yet afford all these conveniences, but he hoped that mass production would soon bring them into Japanese homes. Like many modernists, he believed that the building industry could emulate the automobile industry. He believed that "in the face of the current shortage of four million houses in our country, industrialization is the only far-reaching solution". The name of the project, PREMOS, was an acronym: "pre" for "prefabrication"; "M" for Maekawa; "O" for Kaoru Ono, his structural engineer and professor at Tokyo University; and finally "S" for Sani'in Manufacturing Company. The first two PREMOS units were completed in 1946. The earliest full-scale version was known as "model #7". While maintaining the living unit of the tatami mat as the basic element for these minimal 52-square-metre units, Maekawa incorporated a system of self-supporting honeycomb panels covered by plywood sheeting and

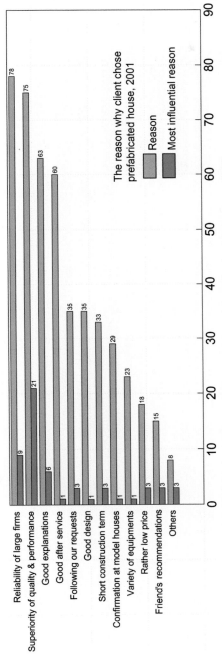

Figure 5.3. Factors motivating clients to buy a prefabricated house (according to Matsumura, 2008).

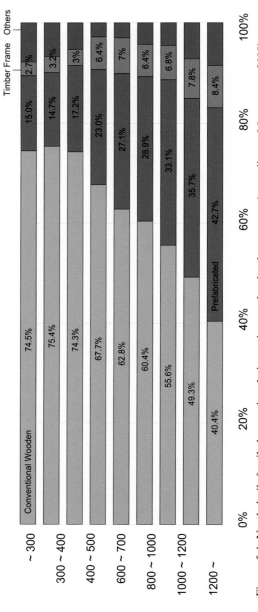

Figure 5.4. Newly built family houses in relation to income levels of customers (according to Matsumura, 2008).

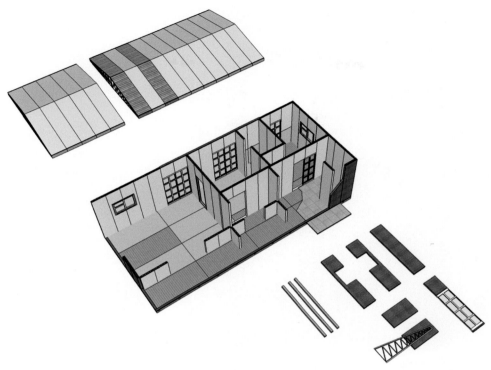

Figure 5.5. PREMOS Home. In postwar Japan more than 1000 units were produced. Graphical representation based on Reynolds (2001) and Matsukuma et al. (2006).

shallow wood trusses to support the roof. By 1952, the joint venture produced more than 1000 units in several variations but all based on the major ideas of "model #7" (Reynolds, 2001; Matsukuma et al., 2006).

5.1.8 Sekisui Heim's M1

Dr. Kazuhiko Ohno developed the legendary M1 building system (Figures 5.6 to 5.8) in 1968 as part of his doctoral thesis at the University of Tokyo. As a result of the expansion of the Tokyo metropolitan area and increased demand for affordable housing in the suburban area, the concept was soon adopted and brought into large-scale production by Sekisui Heim. The system showed flexibility and simplicity in terms of design and had huge potential to be mass produced in an SE/factory environment due to a design oriented towards manufacturability. Through the success of the M1 system in the 1970s, the Sekisui Heim had reached an enormous – for that time – annual production of more than 3000 buildings. The company reinvested a large amounts of the money earned with the M1 in the development of more advanced, automated, and robotic manufacturing technology and thus triggered a performance multiplication effect (PME). The M1 is also viewed as the origin of today's advanced and much more complex and customized/personalized model versions from Sekisui Heim.

Figure 5.6. Sekisui Heim's M1: Visualization, front view.

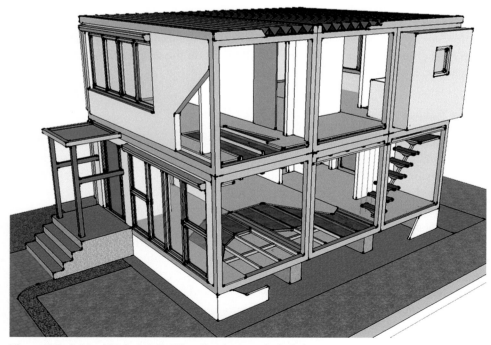

Figure 5.7. Sekisui Heim's M1: Visualization, exploded view.

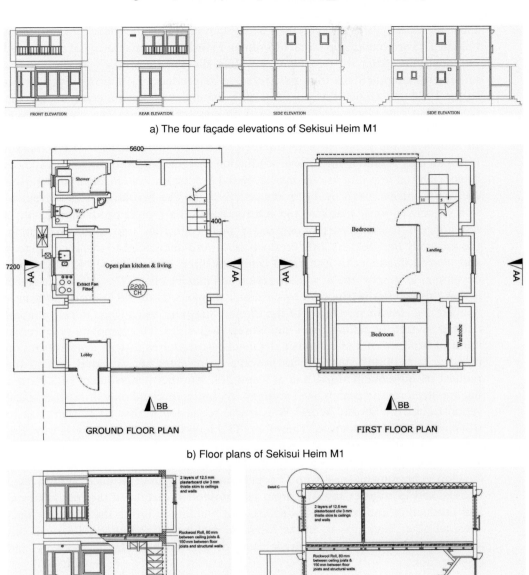

a) The four façade elevations of Sekisui Heim M1

b) Floor plans of Sekisui Heim M1

c) A cross section detail of Sekisui Heim M1 -AA

d) A cross section detail of Sekisui Heim M1 -BB

Figure 5.8. Sekisui Heim's M1: Elevations, floor plans, and sectional drawings. (Drawings by W. Pan)

5.1.9 From Japan's Traditional Organizational Culture towards TPS and Toyota Home

Another milestone in the evolution of mass customized building production was the legendary Toyota Production System (TPS; for further information see, for example Ohno, 1978; Shingo, 1982 and Fujimoto 1999; the principles and ideas behind TPS are outlined in detail in **Volume 1**) for the manufacturing of steel-based building units. After the Second World War, the Toyota Motor Corporation was initially seeking methods to rapidly increase its productivity. During several visits to the Ford and General Motors factories, managers at Toyota came to the conclusion that a production concept based on mass and variation production would never achieve large success, especially under Japanese conditions (Ohno, 1978). In their eyes, the ability to adjust quickly to the frequent changes of the market in Japan was essential for a new production system. Under these circumstances Toyota started to invent its own market-based production system, tailored to Japanese requirements: the Toyota Production System. The revolution was the extension of conventional material and information flows ("push production") into a new concept, based on current demands ("pull production"). In "pull" production, the assembly line delivers only products that were demanded to avoid extra stock and overproduction. An integrated communication system called "Kanban" was developed to support the new information and material flow. The important aspect is that this process is triggered by customer demand. Thus the factories' output is "pulled" by customers, instead of the former "pushing" style, when the output was defined by factory management and storage capacities. The complete synchronization of production and customer demands also requires a strict synchronization between factory and suppliers, as previous work steps are executed only at the request of subsequent steps, and JIT and JIS. Another achievement of TPS was the application of a "zero-waste"-strategy. As Japan received only little economical support after the Second World War, it was requested to find an efficient way to work with existing resources. Therefore, TPS identifies seven major waste producers. The most common waste generator, according to TPS, was overproduction: a product delivered without a customer's demand. In the 1970s, Toyota started to develop its housing business with Toyota Home and began to produce prefabricated houses, and to transfer the TPS from its automotive section to the industrialized and production line–based manufacturing of buildings. During the following decades, all other main players in the Japanese prefabrication industry followed this newly set trend, installing the basic ideas of TPS in their plants, products, and organizations.

5.1.10 Automated and Robotized Production as Sales Argument

An important step in the evolution of prefabrication was the gradual launch of a production line–based, highly automated, and robotic production through major prefabrication companies such as Sekisui House, Daiwa House, Sekisui Heim, and Toyota Home in the 1970s. This also occurred during the threat of potential shortages of skilled and trained workers during a phase of rapid economic growth. The adopted factory organization had many similarities to automotive production (e.g., Toyota Motor's flexible body line [FBL]; see also **Section 4.3.1** in **Volume 1**) and demonstrated that an efficient line production could be reached with the utilization of a "chassis" and "platforms" (see also **Volume 1**, **Chapter 4** for more information on chassis, platform and frame and infill [F&I] Strategies). The conversion of on-site production into off-site production with preassembling also allowed reducing dirty, difficult, and risky work at the construction site. Here it must be mentioned that Japanese people in general have a very positive attitude towards automation, robotics, and technology. Several historical occurrences have shaped a unique view of advanced technologies. For example, in the 16th century the Japanese developed their own types of timepieces ("wadokei"; see, e.g., Yoshida et al., 2005) that allowed people to adjust time measurement to their individual rhythm of work. Later the "ningio karakuri," which can be seen as the predecessors of modern automation and robotics in Japan, had been popularized as mechanized toys for entertainment (Wißnet, 2007). Automation and robotics in Japan gradually gained the image of being designed to serve people – and not the other way around. Also during times of extensive automation in the 20th century, companies shifted people into their service and development sections rather than laying them off. Concerning upcoming challenges as the demographics change and population declines in Japan, robotics (service robotics) and personal assistance technologies are widely accepted as a potential solution. In Japan, the fact that a house is fabricated using automation, robotics, and other advanced technologies has an extremely positive influence on the image of the producing companies and their products. Moreover, the steady success and in turn steadily increased automation degree of the main prefabrication companies such as Sekisui House, Daiwa House, Sekisui Heim, and Toyota have been followed by a clearly recognizable increase in their quality and performance. After decades of reliable products and services with very good performance, the Japanese have now developed a strong trust in their prefabrication companies.

5.1.11 Sekisui Heim – ERP Systems for the Control of Increasing Complexity

After the successful application of the TPS in prefabrication by Toyota, Sekisui Heim followed Toyota and adapted and refined the TPS for their business. In the 1980s, Sekisui then came up with another essential innovation which was soon adopted also by other prefabrication companies: the parent company Sekisui Chemical developed an innovative computer-based enterprise resource planning (ERP) system for controlling production and logistics flow. This ERP system was subsequently transferred to Sekisui Chemical's subsections. In the housing section, this ERP system laid the foundation for HAPPS (Heim Automated Parts Pickup System; for further information see Furuse & Katano, 2006). The system translates floor plan and design requirements of architects and customers directly into production plans, work sequences and manufacturing information. HAPPS was continuously refined since then and today it ensures successful communication between suppliers, work steps on different sections, timing, and feeding of the 400-metre assembly line. Therefore, HAPPS chooses approximately 30,000 parts out of a possible 300,000 listed items for one building, scheduling and arranging them JIT and JIS for production and assembly.

5.1.12 Timeline of Evolution of Prefabrication in Japan

Before 1900	*Traditional Japanese timber construction*: Today's prefabrication industry in Japan has its origins in traditional Japanese timber construction. Traditional Japanese timber construction can be considered as an early example of high-level prefabrication in the building industry. The Japanese tradition is closely related to a strong predilection for order, standardization, and systematization (e.g., Ken, a 1:2 relation proportion and measurement system). Further, Tatami mat layouts follow strict grids and order systems (see, e.g., Osamu, 1994). Typically having an edge length of 85/170 cm Tatami mats can be combined in many variations to shape the room's dimension, which will always result in an exact number of mats. Standardization was a necessity here as the mats were continuously changed between rooms, according to their usage. Today it is still common for room size to be expressed by the number of tatami mats rather than square metres. This standardization results in a particular multilevel grid that can be found not only in the building's footprint but also in its elevations as well as its decorative and built-in parts, such as religious corners, wardrobes, or shoji screens (traditional Japanese sliding doors). This predilection for standardization, originating in traditional Japanese timber construction, created a prosperous environment for the thriving of the prefabrication industry (Linner & Bock, 2012a).
1930–1952	*K. Mayekawa – PREMOS*: The initiation of large-scale factory-based mass production of buildings began in Japan with Kunio Maekawa in the late 1930s and his PREMOS Home. PREMOS not only accommodated inhabitants' resting, eating, and sleeping, but also facilitated cooking and cleaning, adequate light, ventilation, and heat. The name of the project "PREMOS" was an acronym: "pre" for "prefabrication"; "M" for Maekawa; "O" for Kaoru Ono, his structural engineer and professor at Tokyo University; and finally "S" for Sani'in Manufacturing Company. The first two PREMOS units were completed in 1946 and by 1952 the joint venture produced more than 1000 units in several variations. The tatami mat served as a reference for the modularization of building kits. (For further information see Reynolds, 2001 and Matsukuma et al., 2006.)
1945	*Supply of buildings after disasters and wars*: Japan has been frequently hit by large natural disasters – earthquakes (Kanto earthquake, Hanshin earthquake), tsunamis, typhoons, and big fires (e.g., in Tokyo) – as well as the Second World War and the bubble economy 1980s/1990s. After each of these incidents, enormous numbers of houses needed to be built in a very short time. Systematization and prefabrication always served as the basis to fulfil these demands. After the 2011 earthquake (Tohoku earthquake and tsunami causing the Fukushima incident, etc.) the prefab industry was even paid by the government enormous sums of money to quickly supply temporary and permanent shelters for the earthquake-stricken areas.
1960s	*Metabolism – 1960s*: During the 1960s, the metabolism movement introduced visions of modular, scalable/adaptable, and industrially producible mega structures: helix city (Kurokawa), functional building Kits (Ekuan Kenji), capsule systems (Kurokawa), stratiform structure systems (Kikutake), and modular condominiums (Izosaki). For example, in the erection of his capsule towers with, more or less, uniform capsules, Kisho Kurokawa saw the modularity of this metabolistic mega structure as only the first step and envisioned that once a critical mass is demanded, industrialized and production line–based mass customization of the builsing modules/capsules is possible. Sekisui Heim later indeed started to manufacture individual capsule-like units on a production line in a mass production like manner. (For further information, see Kawazoe, 1960/2011 and Hirose et al., 2011.)
1960s	*Intensive promotion of prefabricated steel buildings was pushed forward by the Japanese Prefabricated Steel House Institute*, an institution backed by the steel industry (Kozai Club: Kawasaki Steel, Kobe Steel, Nippon Steel, NKK, Nisshin Steel, Sumimoto Metal). The institute conducted intensive research, built prototype buildings, set up standards and planning guidelines, connected the players, and trained people.

1970/1980s	*Transformation of Japanese joining system into precut technology*: The traditional Japanese house is built up by a variety of wooden column and beam parts, which are only fit together (interlocking technology) and not joined or nailed in a fixed way (see Figure 5.9) so that the joined elements could maintain a certain degree of flexibility (see, e.g., Bock et al., 2011). Many kinds and variations of this joining system have been developed to date. However, the main idea behind the joining system is that it works without steel or metal bolts – only wooden elements are used as subelements. Joining is done by interlocking the wooden parts through complex shaped cuts and sometimes by fastening them with a kind of wooden insert nail (komisen). The joints are, on one hand, precise enough not to separate and on the other hand, guarantee a certain slackness. Distributed over the whole building's structure, a multitude of such joints give the building a controlled flexibility should external forces be applied to the structure (wind, earthquakes).
	The traditional Japanese house erected by the Daikusan (carpenter) suffered from inconsistent quality due to rising labour costs. With traditional Japanese wooden precut technology, parts and joining systems demanded for an individual, customized building could be produced cheaply and on demand by computerized numerical control (CNC) machines (Figure 5.10B). Figure 5.10A shows a set of precut post and beam elements. The general form of the joint's element changed from rectangular to round cutouts, as those could be fabricated faster and more precisely by machine-guided rotating cutter heads.
	Automated precut manufacturing lines were built to perform the cutting process, measure the accuracy and quality produced elements, and sequence and sort the elements.
1970s	*Prefabrication of subunits*: The construction of plumbed rooms on construction sites often creates an organizational bottleneck, as a multitude of trades must be coordinated within a limited space. The prefabrication of plumbing units, however, can take place off-site, in parallel, and without friction to the conventional construction sequence. In the 1970s, prefabricated plumbing units were used mainly for hotels, hospitals, and nursing homes, and (especially in Japan) frequently used in high-rise construction. Today Japan's use of prefabricated plumbing units is no longer limited to a specific building type. Toto and Inax manufacture customized plumbing units on demand for any building type on their production lines. Major Japanese prefabrication companies rely on these units and integrate them as a kind of subassembly into the buildings they manufacture.
1970s	*From Sekisui Heim's M1 to large-scale prefabrication*: In 1968, Kazuhiko Ono developed, as part of his doctoral thesis at Tokyo University, the legendary M1 system of Sekisui Heim. This three-dimensional modular kit was famous for its genius simplicity. It reduced the complexity to allow industrial line–based production. The M1 was a prototype for merging multiple qualities, designs, and production aspects. The "units", based on steel frames, were perfectly suited to the industrial production and even with the low number of components could generate a variety of possible solutions for the customer. In the 1970s, the M1 reached an annual and steady production of more than 3000 units per annum, allowing investment in advanced automation (Linner & Bock, 2012a).
1970s	*Plastics versus steel*: Sekisui Heim initially intended to mass produce a version of Kazuhiko Ono's unit-based housing kit with plastic as the main material for structural elements and thus push Sekisui's competency and capacity in plastics manufacturing. However, after the Korean War and continuing tensions in Asia, Japan, in cooperation with the United States, promoted and supported the use of steel to build up a strong steel industry in Japan. The use of steel in construction therefore became more advantageous than the use of plastic and Sekisui Heim changed its plans.
1970s	*Change towards high-quality products*: Misawa Homes was the first company that introduced a "Home Guarantee System" (1962) and an "After Sales and Maintenance Services System" (1972). All other major prefabrication companies followed this strategy later in the 1970s.
1980s	*Robot boom of the 1980s*: After the first experiments with robots in the 1950s and the introduction of automation and robot technology in many industries during the 1960s, the world experienced its first robot boom during the 1970s and Japan became a leading nation both in robot engineering as well as in robot application/utilization in almost all industries

(continued)

(continued)

	(see also **Volume 1**). Similar to many other companies, Japanese prefabrication companies (most of them had roots in other, more advanced industries) during that time implemented this technology into their factories. In 1985, Toyota introduced the flexible body line (FBL) equipped with a multitude of welding robots and thus introduced fully automatic car body assembly (see also **Volume 1**) in the automotive industry. Interestingly, a few years later, Sekisui Heim introduced a very similar automatic line to assemble and weld its three-dimensional steel framed units.
1980s	*Introduction of TPS elements*: In the 1970s, Toyota developed its Toyota Production System (TPS), revolutionizing manufacturing organization, introducing JIT, demand-oriented material flow (see also **Volume 1**). In the 1980s Toyota started to develop its housing business with Toyota Home and transferred the TPS from its automotive section to the line-based manufacturing of buildings. During the following decades, all other main players of the prefabrication industry followed this newly set trend, implementing the basic ideas of TPS in their plants, products, and organizational structures.
1984	*Introduction of Expert Systems*: Sekisui Chemical Co. Ltd. was the first company in Japan to develop plastic moulds (1947). Today approximately up to 50% of its revenue comes from Sekisui Heim, its housing division created in 1971. Sekisui Heim is famous for its industrialized "Unit Method", which relies on HAPPS (Heim Automated Parts Pickup System) introduced 1984 by ISAC. ISAC is a corporation financed by Sekisui Chemical to develop expert systems and applications that could be used in synergy with its housing division and other Sekisui production enterprises. HAPPS is a parameter-based system supporting the whole workflow: customization, planning, receipt of order, logistics, fabrication, and delivery. It helps to generate parts, component structures, and parts lists from computer-aided design (CAD) floor plans (parts explosion). From this, HAPPS allows to automatically generate logistics schedules, work fows and machine control instructions (Furuse & Katano, 2006). Following Sekisui Heim's approach, other prefabrication companies later introduced similar systems. However, today Sekisui Heim still has one of the world's most advanced manufacturing control systems in the building industry, allowing more than 90% of all design- and parts-related information to be directly translated into production and assembly operations. Furthermore, Sekisui already also uses the system to manage delivered building products over time in terms of customer relations, maintenance, upgrade offers, rearrangement, and deconstruction/recustomization (Linner & Bock, 2009).
1990s	*Zero-waste and recustomization*: In contrast to conventional construction, throughout prefabrication in structured environments (SEs) in the factory, minimum waste is generated. Material streams are completely transparent and controllable in the SE and remaining waste is fastidiously sorted for reuse. Both Sekisui Heim and Toyota Home operate "zero-waste-factories". Furthermore, all buildings from Sekisui Heim, for example, can be accepted as trade-ins for a new Sekisui Heim buildings. Sekisui Heim takes back its units, disassembles them, removes finishings, and equips them on the production line with new finishing (recustomization).
2004	*Tron House 2*: Toyota Papi (Tron House 2) is an experimental prototype of a next-generation prefabricated building developed by Prof. Ken Sakamura designed and developed in cooperation with Toyota Home, as a new intelligent home with novel performance elements based on the extensive integration of advanced technologies in the form of distributed sensors, micro systems and mechatronics. The main objective of this project was to create and realize an environment-friendly, energy-saving, intelligent house design, in which the latest computing technologies, developed by Sakamura's T-Engine project, could be tested and furthered. The whole building is based on Toyota Home's steel frame units (Linner & Bock, 2012b).
2006–2009	*Zero-carbon emission houses*: Because of the steady increase in worldwide resource consumption and CO_2 emissions, all major Japanese prefabrication companies focussed on research concerning building technologies and materials that allow for more eco-friendly buildings. To date, zero carbon emission and energy efficiency has remained one of the main topics in the Japanese prefabrication industry. All companies have introduced

advanced features for intelligent CO_2 emission reduction and energy consumption in their regular model series. At the same time, following a strategy announced by the Japanese government to promote 100- to 250-year durability of buildings to save resources, prefabrication companies increasingly emphasize the potential durability of their building's basic/bearing structures.

2010   *Start of business operation in other countries*: To address the stagnation and decline of the housing market in Japan, Japanese prefabrication companies gradually oriented themselves to overseas markets. Sekisui Heim, for example, erected a factory in Southeast Asia whereas Daiwa House extended its operations to mainland China.

2011   *Robot technology and buildings*: Daiwa House cooperates with robot manufacturers such as Cyberdyne in introducing robot technology into daily life and for assisting elderly people in an ageing society.

2011   *Advanced disaster management*: In response to the Tohoku earthquake and tsunami incident, Japan's prefab industry – which already provides outstanding quality in terms of earthquake resistance – once more started to pushed research in improved structural resistance of their buildings. Furthermore, to be able to support people efficiently during future major disasters, Daiwa House presented its EDV-1 disaster relief unit, which will be marketed through Daiwa's Leasing Division (Daiwa Lease) in the future. The EDV-1 is a compact high-tech temporary shelter that provides all life-critical functions during the first weeks or even months following a disaster. (For further information see Daiwalease, 2013 and also system outline at the end of this chapter in Section 5.5.2.)

2012   *Daiwa House acquires general contractor Fujita*: Fujita had already built up businesses in emerging markets and Daiwa House wanted to use Fujita to obtain a threshold in those markets. As the Japanese prefabrication market incurred stagnating export to other countries, it was an important issue among prefabrication companies in Japan.

2013   *Service innovation*: On the basis of the already deployed large-scale industrialization, companies more and more shifted their focus from delivering products towards physical and digital services that were related to the buildings utilization phase. Major prefabrication companies therefore introduced various new divisions or spin-offs.

2014   *Towards prefabricated settlements and developer approach*: Panasonic cooperates with other companies to develop and build whole (technologically advanced and eco-friendly) settlement. The the settlement will incorporate latest building and smart grid technology, novel mobility concepts, assistance technology and other sensor and micro system technology-based service concepts. Several major prefabrication companies shift more and more to a similar developer and services oriented approach.

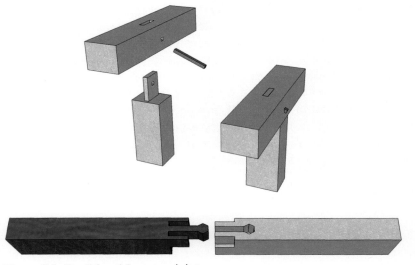

Figure 5.9. Traditional Japanese joints.

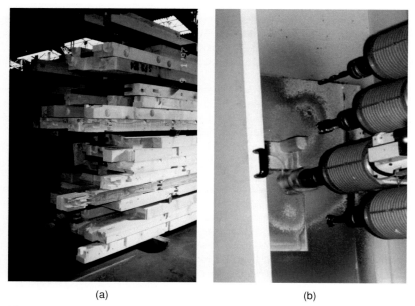

(a)   (b)

Figure 5.10. Precut technology. A) Set of precut post and beam elements. B) CNC-controlled rotating cutter heads.

5.2 Robot-Oriented Design and Management Strategies used in the Japanese Prefabrication Industry

5.2.1 The Idea of Robot-Oriented Design and Management

In **Volume 1**, strategies and tools were presented that allow for the efficient introduction of automation and robot technology in the construction industry. It was argued that automation and robot technology elements introduced in construction, as in any other industry, cannot be viewed as stand-alone entities. Strong complementarities exist among product structures/designs, manufacturing technology, information technology, and management and organization. It was shown that the introduction of automation and robot technology can efficiently and successfully be accomplished once all business operations and, in particular, design and management are adapted to the new manufacturing technology. The success and large scale of the Japanese prefabrication industry is attributable to the consequent adaptation of all business operations towards the aim of systemizing and automating processes. Neither in any other prefabrication industry around the world, nor in automated/robotic on-site factories (discussed in detail in **Volume 4**), has such extensive adaptation taken place. Therefore, in the following section, key aspects of this co-adaptation are identified and explained. Detailed explanations of the concepts and tools behind co-adaptation strategies are given in **Volume 1.**

5.2.2 Complementarity as a Key Element in the Success of Automated Prefabrication in Japan

As discussed in **Volume 1**, in the manufacturing industry, in general, the integration of products, organization, informational aspects, and machine technology is the key to automated product generation and the steady increase of efficiency, product quality, and product complexity. The integration of these factors is also the basis for the customization, or personalization of products, on the basis of automated manufacturing systems by means of product modularity (integrated user and producer kits), organizational setups, and computer-aided closed informational chains. It can create systems that directly link demand and the customer (without the need to convert digital into analog/manual instructions and back) to the manufacturing machines that actually generate the product. In the Japanese prefabrication industry, the main ideas of the theory of complementarities (Milgrom & Roberts, 1990; for further information see **Volume 1**) are realized par excellence. A major reason for the openness and willingness of companies to adapt their complete business operation to new, advanced forms of manufacturing can be seen in the fact that the industry (as well as and most companies, see Table 5.1) is very young and that most companies have no tradition in building construction, but rather originate from large companies with a strong foothold in other more advanced industries (e.g., Sekisui Chemicals, Panasonic, Toyota, Asahi, etc.).

A co-adaptation of novel automation and robotic base manufacturing technology as well as the manufacturing of optimized product structures (clearly not based on conventional design strategies) in the Japanese LSP industries have enabled the generation of buildings of a generally higher quality and higher degree of individuality than those produced, for example, by the locally and tradition-oriented German LSP industry (see also Section 4.1.2). The Japanese LSP industry demonstrates that a large-scale, high-degree of automation and the abandonment of tradition are – contrary to the commonly accepted thinking in architecture and construction industry – actually the drivers and enablers of productivity and high value-added, personalised (high-tech) building "products" and services. In addition, employment structure (demand of skilled knowledge workers) and the extraordinarily high salary levels (see Section 5.1.1) in such an industry meet – in contrast to conventional construction industry – the requirements of a high-wage and leading industrial nation where workforce tends to have a high level of education and skills.

One particular characteristic of the Japanese LSP industry that ought to be emphasized is the OEM-like integration structure that almost all major prefabrication companies adopted to minimise lead times for their buildings (products) and to efficiently assemble parts, components, and modules in their off-site SEs. Although German LSP companies such as Weber Haus and Baufritz achieve a high degree of automation and excellent quality control in their factories, because of the missing concentration of large companies, the local orientation of integrating companies and suppliers, and the comparably low output, they do not achieve the build-up of the OEM-like integration structure on which the Japanese LSP industry is based today. As explained in **Volume 1**, an OEM-like integration structure allows for the economically efficient pre-manufacturing of components and other high-level parts and modules by Tier-n suppliers and is a prerequisite for the production line–based and highly systemized, automated or even robotized final assembly in a one-piece flow (OPF) manner.

5.2.3 Robotic Logistics-Oriented Design

In Japanese LSP in general, the product is designed to optimize material flow both in terms of factory external and internal logistics (see **Volume 1, Section 4.3.3,** for a detailed outline of logistics strategies and technologies relevant in automated/robotic construction). Analysis and comparison of various prefabrication strategies used in the Japanese prefabrication industry show that steel systems, and in particular three-dimensional space frames (both represent a break from conventional building structures), allow for a consequent and production line–based manufacturing organization (with uninterrupted, optimized flow of material) and thus the highest possible degree of systematization and optimization of manufacturing.

1. In contrast to panelized systems, three-dimensional steel space frames (e.g., called *Units* by Sekisui Heim), concentrate the manufacturing activity on a few units flowing through the factory (e.g., for a building: about 15 units versus 40 to 50 panels that would flow through factories producing panel-based buildings) and thus allow to simplify, focus and optimize the material flow in the factory. In addition, the three-dimensional steel space frames act as a "carrier element" and accurate template (similar to the car body in automotive manufacturing, for example) that can be equipped with parts, components, and modules (e.g., plumbing units) on the production line. Three-dimensional space frames also provide a template for the fast and efficient installation (positioning, alignment, and fixation) of "bulky" high-level components as, for example, interior-wall and facade elements.

2. Panelized systems as, for example, two-dimensional steel frames (e.g., used by Sekisui House, Daiwa House, Pana Home) and two-dimensional wood (e.g., used by Mitsui, Misawa) panels also can be used as a carrier element in the factory/SE; however, as these elements do (unlike the three-dimensional space frames) not represent full rooms or parts of rooms the number of parts and components that can be attached to them is limited. Larger equipment, appliances, furniture, and modules such as plumbing units cannot be carried or taken up by them and also the cannot be used as an accurate template for the positioning, alignment and fixation of higher level components as, for example, walls.

3. Also three-dimensional wooden frames (Sekisui Heim Two-U product line) do not allow for consequent, otptimized organization along one main line from beginning to the end in an SE, as the production of a wooden frame's bottom, sides, and top panels as well as their assembly into a three-dimensional space frame involves a multitude of workshop-like activities, and as machines and end-effectors for wood parts/components processing are (due to component weights and dimensions of wood which do not favour automation; see also **Volume 1, Section 4.4**) less advanced as those for steel parts/components processing. However, once the three-dimensional wooden space frames are set up, applying the finish and equipping them with installation equipment, appliances, and furniture can be done JIT and JIS on the production line as in the final assembly of units based on steel. Therefore, in Sekisui Heim's new factory in Sapporo, three-dimensional wood and steel frames already share one final assembly conveyor belt where final assembly takes place (see also Figure 5.23).

5.2.4 Robotic Assembly–Oriented Design

In Japanese LSP in general, the product structure is synchronized with the composition/definition of stations, end-effectors, and the factory's internal automation ratio.

In particular, the use of steel as the structural basis for prefabricated buildings by many companies (in the form of panels and three-dimensional space frames) has an advantageous influence

- On the use of automated systems and robot technology
- On the application of JIT and JIS throughout the whole OFM process

The use of automated systems and robot technology is simplified by steel as a multitude of systems for automated/robotic positioning, adjusting, and fixation of steel elements are available and the technologies proven (e.g., various welding technologies). Sekisui House, for example, uses robots to assemble, adjust, and weld the bare steel frames. Sekisui Heim uses a three-dimensional system (similar to the automatic car body assembly line, also called flexible body line [FBL] in use at Toyota Motors; for more detailed information see **Volume 1**) to automatically position, adjust, and spot weld the bottom, top, and side elements of the steel units. Because of the stiffness and compactness of steel (compared to wood), the resulting component carriers/assemblies, templates, and the number of parts to be assembled have been reduced, and the accuracy and speed have been enhanced.

5.2.5 Degree of Structuring/Automation of Off-Site and On-Site Environments

Three-dimensional fully equipped units reduce the amount of on-site work, build the basis for systematization and automation in the off-site SE (see Section 5.1.4), and reduce the total lead time. They also allow for high quality standards to be achieved, as most of the work can be conducted in the factory's SE and thus in a controlled and automation-supportive environment by machines and robots (Figure 5.11). The high degree of prefabrication (factory-based work) does not negatively influence the degree of customization/personalization of the buildings, as each unit produced in the SE is equipped in an OPF manner, with individual finishing, installations, and equipment. However, three-dimensional units have the disadvantage of increased cost and effort for the transport from the factory's SE to the site. So far, Sekisui Heim is the only company that has been able to partly compensate for this disadvantage by its factory network (six factories distributed throughout Japan), assigning the manufacturing/final assembly of the units to the factory located nearest the site.

A disadvantage in the Japanese LSP industry is the "exclusiveness" of the systems offered. A building built from three-dimensional steel units (e.g., Sekisui or Toyota) must be built up completely using those units, and the units can only come from this one manufacturer. Also, mixing panels or conventional construction methods (for example) are avoided by the manufacturers. Usually, companies delivering buildings based on three-dimensional space frames make a profit with bath and kitchen units (high amount of installation and equipment) and lose efficiency with other rooms. On the other hand, companies that work with panels face disadvantages concerning rooms with a large amount of equipment as it must be integrated on-site, in a rather unstructured environment, and gain efficiency only wherever relatively simple wall panels can be combined. A combination of three-dimensional units for complex, installation-intensive rooms as kitchens and bathrooms, and

Figure 5.11. Robotic steel frame manufacturing at Sekisui House. Setting with cooperating robots fixtures and jigs in fabrication of steel frames. (Photo: Sekisui House)

panels for other rooms would thus unify the advantages of both systems. Currently, no Japanese LSP company is able to deliver such kits in a materially uniform manner (only Misawa would be able to deliver steel units and wood panels). Another disadvantage of the LSP industry is that on-site assembly processes, although manufacturing off-site in the SE is in most cases highly automated, are still predominantly based on human labour (which, in the case of companies such as Sekisui House/Daiwa House, working with panels, still represents about 50% of the necessary work), which leads to the fact that the consequent systematization and automation approaches applied in the off-site factory environment are not continued on-site as well.

5.2.6 OEM-like Integration Structure

Companies such as Sekisui Heim, Sekisui House, Toyota Home, and Misawa Homes (Hybrid) in particular have altered the building structure as well as the manufacturing process and organizational structure dramatically compared to conventional construction, and state the final integrators in an OEM-like integration structure where subfactories or Tier-n suppliers deliver parts, components, and modules that are themselves preassembled or prefabricated (for more on this strategy see **Volume 1, Section 4.3**). Toto and Inax (both major suppliers of bath equipment) prefabricate plumbing units (especially bath cells) as well as kitchen units and thus serve as Tier-1 supplier supplying high-level components and modules to the OEM. More details on the phenomenon of building module manufacturing by Tier-1 suppliers in the Japanese prefabrication industry are given in Chapter 3.

5.2.7 Modular Coordination

The importance of modular coordination to achieve a synergistic co-adaptation of product structure/design and manufacturing method/equipment in the sense of ROD is outlined in detail in **Volume 1** (in particular **Section 6.5**). Sekisui Heim and Toyota Home both apply a frame and infill (F&I) strategy organized around three-dimensional steel space frames, and most other companies on the basis of (two-dimensional) steel or wood frames (e.g., Sekisui House, Daiwa House). The visible modularity of these units, frames and panels (individual and characterizing for each of the companies outlined in Section 5.4) is the distinct embodiment of the integration and synchronization of a specific set of manufacturing relevant aspects, as, for example, of supply chain aspects (modularity advantageous for OEM-like integration), customer integration aspects (choice, platform strategies), factory and assembly organization (e.g., production line–based SE), machine/robot/end-effector aspects (e.g., necessary working spaces, DOFs, and accuracies), and logistics aspects (factory internal and external logistics systems and their capacity) with the modular structure of the total building and the units.

5.2.8 Control of Variation by Platform- and Same-Parts Strategies

In the automotive industry, platform strategies aim at providing a technical basis to which various attaching parts are mounted (see **Volume 1, Section 4.2**). In a similar way, Japanese prefabrication companies use their steel frames (e.g., Sekisui House) or steel space frames (e.g., Sekisui Heim) as an adjustable platform on which different types of buildings and their individualized descendants are built. It is the common basis for all types and variants allowing stable production processes (for more information on the process stability that can be created by platform strategies, see, e.g., Schmieder, 2005 and Piller, 2006; see also **Volume 1**, **Section 4.2**) despite the ability to customize houses to suit any need. About 10 new Sekisui Heim housing models and about 400 modifications and improvements to already existing models are introduced annually. It is therefore necessary to be able to easily modify the product structures through platform strategies by allowing that individual modules are adjusted, added or improved without having to redesign the platform and other modules. Based on modularity, all major Japanese prefabrication companies have deployed product structures for their houses that can be identified even as open platforms enabling "open" product evolution while at the same time minimizing the impact of engineering/product alterations on (stable) manufacturing processes.

5.2.9 Linking of Customer and Manufacturing System

From the 1990s on, the CAD/CAM field evolved into computer-integrated manufacturing (CIM; for further details see **Volume 1**). The focus was now broader and the idea was that more and more fields, tools, and even business economic issues (computer-aided forecasting or demand planning) could be integrated by computerized systems to form continuous process and information chains. In the following decade, mass customization continued that idea by demonstrating that even the customer (beginning and end point of any product generation) can become part of such processes and information chains. With off-line product configurators, for example, tools that extend the idea of computer integration fully into the customer's needs and activity field were developed by Japanese prefabrication companies, allowing for direct linking of the customer (and his/her activities and demands) to the manufacturing system. According to Milgrom and Roberts (1990), the implementation of a mass customization strategy along with speedy, computer-integrated, and automated production exploits complementarities, and thus the new manufacturing technology not only reduces cost but also adds value to the products. The linking of the customer to the manufacturing system through a mass customization strategy is a key characteristic of the Japanese prefabrication industry.

Off-line Configuration in the Japanese Prefabrication Industry

The configuration process is done by all companies in several steps and is guided by trained customer contact staff who help customers make decisions as quickly as possible. Off-line (i.e., staff-guided) configuration software tool allows choosing from different types and variants and enables adjustment of finishing and details according to individual needs. Before trained staff configures and "co-creates" the building together with the customer using the off-line customization software tool (which allows to translate the resulting configurations via HAPPS directly into work flows and manufacturing equipment control information), the customers are invited for test-living and a systematic (or better: scientific) analysis of anthropometric, physiological, ergonomic, psychological and habitual aspects in their R&D and customer centers (Figure 5.12).

Flexible Degree of Customer Involvement

Throughout the configuration and co-creation process, Japanese prefabrication companies allow their customers to individually choose their preferred degree of customer integration. Yet, the degree of customer involvement still determines the price. Only if the customer selects one of the basic types and chooses a standard floor plan, standard interior finishings, windows, and façade elements, the price of the house will be kept at a minimum. However, the choice is always up to the customer. All companies are able to (and also offer this) prefabricate houses with fully personalized floor plans, features and designs and customer determined or even customer designed individual finishings. Continual process improvements aim at gradually lowering the price impacts of customization. During the development of new models and the marketing of existing ones, it is also typical that customers be invited to visit one of the research and development centres to test various parts of the housing systems for criteria such as usability and accessibility. However, customer needs are not static and must be served continuously. Japanese building manufacturers have realized this and work on continuously extending the degree of customer involvement into two directions: involvement in product development and involvement in product life-cycle related services (Figure 5.13).

In terms of involvement in product development, generally speaking, the companies Sekisui House and Daiwa House are leading by allowing customers to choose from a wide variety of standard parts and designs, which can also be redesigned and reengineered based on individual customers' demands. Sekisui Heim and Toyota Home in general do not offer such a high variety of predefined standard parts, and the customers have fewer opportunities to adjust standard housing types and designs to their individual demands – their very systemized and on three-dimensional steel frames based manufacturing strategy limits involvement basically down to the customer influenced combination of predefined parts, components and modules. However, the degree of customer integration in development issues still determines the price and therfore companies as Sekisui Heim and Daiwa House (that allow customer involvement beyond combination of modules) are still among the most expensive. Research and frequent visits by the authors to various production facilities of all mentioned companies have shown that Sekisui Heim and Toyota Home, on the contrary, have the highest degree of factory prefabrication (work done in the factory accounts for up to 85%) and slightly higher degree of mechanisation, automation and robot technology utilization since the lower degree of customer involvement leads to less variability and thus to more stable manufacturing processes. Furthermore, Japan's building manufacturers gradually extend their focus to the products use phase and in particular services delivered through that use phase which hold the potential for additional value creation possibilities. Companies thus try trying to remain in touch with and serve customer continuously through the whole product life cycle (see also Section 5.5.1). A leading company in this field is Sekisui Heim, which not only offers long-term maintenance service packages and

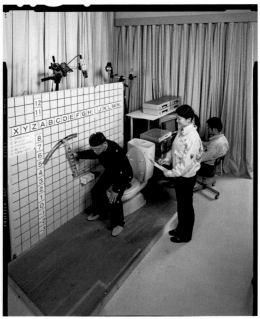

a) Within a testing environment staff of Sekisui House and the customers "co-create".

b) Within a testing environment biomechanical analyses are done with the customer and the outcome is used to configure and optimize the building's design.

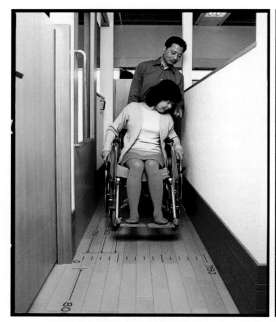

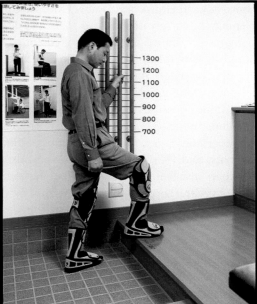

c) Customers experience the use of a wheelchair in later life stages in a test flat and make better decisions concerning room layout configuration in collaboration with company staff.

d) Customers experience the process of ageing by wearing an age simulation suite in a test flat. On basis of this the customers optimize their decisions concerning design options.

Figure 5.12. Customer integration at Sekisui House. (Photos: Sekisui House)

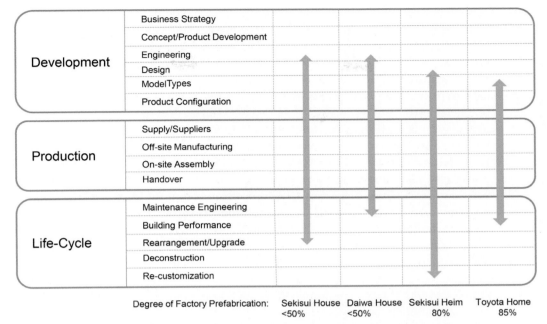

Figure 5.13. Japanese building manufacturers continuously extend the degree of customer involvement into two directions: involvement in product development (direction one) and involvement in product life-cycle related services (direction two).

upgrade services bundled to their housing products, but also allows deconstruction, recustomization, and relocation of its prefabricated buildings.

5.2.10 Innovation and R&D Capability as Key Elements of the Business Strategy

In **Volume 1**, innovation and R&D strategies relevant to automated and robotic production were introduced and discussed. The application of automation and robot technology in construction can be considered as an innovative approach that requires the integration of a multitude of complementary strategic and technological innovations for its full realization. In the Japanese prefabrication industry, innovation and R&D capability play an important role and must be fully compatible with the factory prefabrication approach.

Analysis has also shown that most German LSP companies are relatively old companies (founded between 1900 and 1950) that have not considerably changed the structure of their products (see Chapter 4). This is in contrast to the relatively young Japanese manufacturers (see also Table 5.1) that have conducted a radical change from traditional wood-based construction to radically new product structures such as steel panels, three-dimensional steel frames, and three-dimensional wood units. Apart from the young age of the Japanese LSP industry, the consequential modularization of products and manufacturing systems (as outlined in **Sections 4.2** and **4.3** in **Volume 1** a prerequisite for fast evolution) can be considered as a major driver for the fast development of the Japanese LSP industry. How the Japanese LSP industry is innovating further, and might eventually change from a product-based industry into a service-based industry on the basis of modularization, manufacturing technology, and close and continuous customer relationships over time (see also Section 5.2.9), is outlined Section 5.5 and analysed in detail in Linner & Bock (2012a).

As previously discussed (Section 5.1.2), the intensive R&D concerning the building's performance (earthquake resistance, energy efficiency, etc.) is a direct effect of efficient production processes, creating a PME (see **Volume 1** for a more general explanation of PME and Section 5.1.11 for LSP – specific information) freeing up budget allotment for both reinvestment in better manufacturing technology and in R&D related to the product's features and performance. The Japanese prefabrication industry has thus managed to overcome the in conventional construction usually very low (see **Volume 1, Section 4.1.6**) reinvestment and R&D spending.

The outstanding performance of the products of the Japanese prefabrication industry, bundled with services related to the buildings life-cycle and intense customer "care" (and all this at high but still reasonable cost), is the key performance feature (and selling argument) of the Japanese prefabrication industry. In order to create the outstanding performance and service features that distinguish those building products from competing offers, all housing models and prototypes of new or existing models are subject to continuous improvement, (scientific) experimentation and systematic testing in the companies' R&D centres. The building is treated as a product and testing of its behaviour in response to rain, snow, fires, sound, earthquakes (Figure 5.14), typhoons, tsunamis, ergonomy, physiological and health aspects, psychological aspects and other influential factors is done. In contrast to conventional construction, prefabrication companies apply methods of scientific and systemic product development and usability engineering (prototyping – testing – improving).

5.2.11 Performance Multiplication Effect

As discussed througout **Volume 1,** a performance multiplication effect (PME) refers to the multiplication of productivity or efficiency within an industry, followed by the introduction of advanced machine systems and continuous product improvement. Industries such as ship building, tunnel boring, and automotive and aircraft building have, with the switch from crafts-based to machine-based manufacturing, not only incrementally improved manufacturing performance but actually multiplied it (e.g., by reducing lead time down to one tenth, reducing required human labour down to one tenth, etc.). The achievement of a PME can be seen as the result of an efficient exploitation of complementarities. Analysis has shown that work productivity (output per employee) in the Japanese prefabrication industry clearly increases with the scale of the company and the number of houses built. The work productivity of Japanese LSP companies compared to German LSP companies is enormous (output per employee per year – Japanese prefabrication companies as Sekisui House, Daiwa House: 3.2; Japanese prefabrication companies as Weber Haus, Baufritz: 0.7–0.9; according to 2011 data). The figures become even more impressive considering that the German LSP company with the highest productivity (Kampa Haus) produces houses of relatively low quality and for low-income markets. The quality (and cost)

Figure 5.14. Earthquake resistance testing in an R&D center of Sekisui House. (Photo: Sekisui House)

of a Baufritz house is comparable to the average quality of Japanese prefabricated houses. The business volume (turnover) of Japanese LSP companies, which for each company is in the billions (euros), clearly shows the scale and impact of the Japanese prefabrication industry. All in all, it can be said that products of the highest quality are manufactured in the Japanese prefabrication industry with minimal inputs – both minimal input of human labour and input of material – and thus with high labor and resource productivity. As outlined in **Volume 1, Section 5.1**, increasing productivity is an indicator of technological change. The systematization of the construction process in an off-site SE, and the successfully application of mechanization, automation and robotics in those environments is the backbone of the Japanese LSP industry. Furthermore, it is remarkable that the average salary of employees of Sekisui House or Daiwa House continuously grew since the 1970s and is now more than €70,000 per year after taxes (Nikkei BP, 2009 and also Section 5.1.1). This is more than double the salary of, for example, an electrician in Germany. The average salary at Sekisui House and Daiwa House is on the same level as the average salary in the Japanese automotive industry (Honda, Toyota) or even slightly above it (Nissan, Subaru). However, it has also to be remarked, that both companies do not include subcontractors in their salary statistics – this is probably owed to the war for talents among Japanese high-tech industries. Here Japanese prefabrication companies are in direct competition to companies from industries such as the automotive industry and need to make any attempt to attract young and skilled engineers, who are likely to choose those companies that offer the highest salaries.

5.3 The Manufacturing Process

Basic knowledge about manufacturing processes, manufacturing related modularity and advanced manufacturing equipment used in general manufacturing industry as well as in automated/robotic construction was introduced and explained in **Volume 1** (in particular in **Chapter 4**). The following sections build on this knowledge and extend it in order to be able to understand the manufacturing strategies and processes applied by Japanese prefabrication companies.

Sekisui Heim, Toyota Home, and Misawa Homes (Hybrid) break down a building into three-dimensional units. Those units are realized on the basis of a three-dimensional steel space frame that provides the bearing (steel) structure of the building and can be sent on a production line where it can be almost fully equipped with technical installation, finishing, kitchen, bathrooms (plumbing units), and appliances. Misawa is thus able to shift 90% of all necessary work to complete a building into the SE of a factory, Toyota Home up to 85%, and Sekisui Heim up to 80%. The manufacturing process of Sekisui Heim is the only one in which, from the assembly of steel profiles to the packaging of the finished modules, material flow and assembly operations are organized without interruption or buffers along a stringent production line in an SE and that can therefore be considered a complete and strictly organized flow manufacturing system. Specifically in the first half of the manufacturing process on the production line, automated machines and robot systems assist or perform assembly, adjustment, and fixation operations. Sekisui Heim has the most strictly organized and most automated manufacturing process in the Japanese prefabrication industry.

Sekisui House and Daiwa House, in contrast to the aforementioned companies, break down a building into steel frames (panels). These steel frames are then equipped with interior and exterior finishes (interior walls, facade elements, insulation) and delivered along with other parts and components to the site as a kit. The amount of work shifted to the SE in a factory for these system amounts to only about 50%. In the case of Sekisui House, the manufacturing of the frame kit is highly structured and organized in a flow-line–like/chain-like manner and conducted with the use of robots for assembly and welding operations. In the case of Daiwa House, the factory process is also highly structured; however, the material flow is not rigidly organized as a chain of workstations (only flow-line–like general direction of flow of material is achieved), and certain assembly and welding operations in some of its factories are relatively labour intensive. As the system and organization of Daiwa House are similar to those of Sekisui House and advanced machine technology does not play a role in their organization, the manufacturing process of Daiwa House will not be explained in detail in this section. However, it has to be mentioned that also Daiwa Houes uses advanced technology, off-line configuration and automation and robot technology at certain workstations. Daiwa House is more an innovator in management and business methods and is well known for its developer approach (see also Section 5.1.2) and moreover for an excellent deployment of management and worker training programs (not only for the factory workforce but also for local workers and subcontractors that are allowed to assemble Daiwa House buildings in a franchise-like manner), creating a highly organized and skilled workforce (for factory as well as on-site), which ensures a high quality of assembled buildings.

Besides the aforementioned steel-based systems, some Japanese prefabrication companies also work with wood-based building kits that are prefabricated in the SE of a factory. Sekisui Heim, for example, produces three-dimensional wood-based units in the factory (Two-U Home) with a degree of work shifted to the SE of approximately 60–80%. Similar to the manufacturing of the three-dimensional steel units in the factory, wood units travel along a production line and are equipped with technical installations, finishings, and appliances. However, the process of assembling the wood frame unit is not as rigidly organized and automated as the steel unit assembly and also on the construction site more installation and finishing work is required. Companies such as Misawa and Mitsui have product lines that fully break down a building into wood panels and are then manufactured in a flow-line–like, or as in the case of Misawa, partly chain or production line–like manner in the factory. Aside from their panelized systems, Misawa, Mitsui and also Sekisui House offer precut kits consisting of columns, beams, and other wood parts with precut joints. Those precut kits can be bought by construction firms, contractors and also individual home owners as basic components to build semi-prefabricated buildings on-site. All in all, with the current rise of awareness for ecological issues in Japanese society, it is expected that the demand for wood-based buildings and precut kits will rise in the short and midterm and that rigid organization, continuous line–based material flow, and automation and robot technology from steel-based prefabrication will be transferred more and more also to wood-based manufacturing. For example, Sekisui Heim's Hokkaido based factory already unifies steel- and wood-based manufacturing under one roof and the three-dimensional units share already a common final assembly production line. The factory is highly modular and the section for the assembly of the wood units can be extended in case of a rising demand for wood-based buildings.

5.3.1 Product Variety and Types of Prefabrication

Not only is the amount of prefabricated buildings and components in the Japanese housing industry enormous, but also the variety of available system types. Specifically plumbing unit systems, wooden post beam systems (precut kits), and wooden panel systems are not closed systems but can be combined with other prefabricated components or with conventionally build building sections. Table 5.4 outlines the key types of product structures used in prefabrication in Japan.

5.3.2 Production Process Explained by Sekisui's and Toyota's Unit Method

In this section the production strategy and processes are explained exemplarily in detail by a case study of Sekisui Heim's production process (see also Linner & Bock 2012a; the following section is a rewritten, expanded, and improved version of a section of this journal article).

Table 5.4. *Types of prefabrication used by the Japanese housing prefabrication industry*

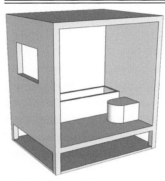

a. Plumbing modules

b. Wooden frame and component kits

c. Wooden panels (small, medium, large)

d. Steel frames combined with panels

e. Fully panelized steel kits

f. Steel-based units

g. Wood-based units

Sekisui Heim breaks down a unique family house into 10 to 15 units, each one finished up to 85% in the SE of a factory. Into these units, a multitude of other three-dimensional subsystems from various Tier-1 suppliers (e.g., bath and kitchen modules; see for details Chapter 3) are integrated in production line–based SE of a factory. This production method shows many analogies to automotive production, for instance, that a continuous production line becomes possible through the introduction of a three-dimensional steel frame used as the "chassis" (serving as template and component carrier). This "chassis" support structure is sent through the SE with a speed of 1.4 metres per minute, passing by more than 45 workstations on a conveyor belt with a length of approximately 400 metres. During this process, it is – partly automatically – equipped with modules, components, and subsystems. The "chassis" is sequentially and three-dimensionally finished from all sides with components (e.g., wall panels) either being supplied JIT or having been produced in parallel processes on different floors or in different sections of the factory.

Advanced Supply Chain Management

Innovative strategies for ERP and logistics determine the efficiency of the resource flow by controlling the processes that feed and integrate resources (input factors such as material, human beings, and tools/equipment) into the company's value creation system. JIT and JIS industrialized production lowers input resources and inventory (see also **Volume 1, Section 4.3.1**). It has the potential to counterbalance additional costs of product customization/personalization. The "Kanban" system introduced by Toyota and Taichi Ohno with the TPS (Ohno, 1988) can be considered the precursor and activator of innovative OPF systems aimed at increasing individualization of production outputs. Kanban and TPS have also been transferred from Toyota's automotive sector to Toyota's housing sector. Here, it was later also adopted and improved by other major prefabrication companies such as Sekisui Heim. Advanced BIM is a precondition for the successful use of tools like Sekisui Heim's HAPPS (see also Furuse & Katano, 2006) that allows a translation up to 95% of CAD and with the user co-created input data, into logistics, production, and assembly information.

Main characteristics of the supply chain management system:

1. *Automated component selection*: The evolution of Sekisui Heim into a highly productive company was initially enabled by its advanced IT-based ERP system called HAPPS. As explained previously, houses are made of 10 to 15 steel frame units, all individually finished according to customer demands. This means that each unit, prefabricated in the factory, is different. Therefore, it is a complex process to select and pick up roughly 30,000 components correctly for each house – out of approximately 300,000 available – and feed them to the production line in the correct sequence.

2. *Automated task and production scheduling*: HAPPS is a parameter-based system supporting the entire workflow: configuration, planning, receipt of order, logistics, fabrication, delivery, maintenance/servicing and recustomization. It helps to generate parts, component structures, and parts lists from the (user

co-created and in CAD-represented) building configurations. Based on the information generated from the CAD models, logistics, material flow and work procedures are generated and controlled to a high degree automatically.

3. *One-Piece Flow*: OPF refers to a production method in which each entity or unit, moving along the main production line, is allowed to be different (see also **Volume 1**, **Sections 4.3.1** and **7.5.2**). It is particularly important in the industrialized housing industry that each steel frame unit can be finished on the production line with a high percentage of individual designs and floor plans. Today, both Toyota Home and Sekisui Heim have adopted OPF principles.

Line-Based Production and Assembly

The factory organization of both Sekisui and Toyota are based on assembly line production where the moving steel frame units are customized according to floor plans, functionality, technical infill, and finishing demanded by individual customers. Subcomponents (e.g., walls that are then later on installed to the steel frames on the main production line as can be seen in Figure 5.15) are fabricated in parallel processes in sub-lines on various floors.

Main characteristics of the production and assembly system:

1. *Automated steel frame production*: One of the basic features is the automated assembling and welding station. Ceiling elements, flooring elements, and columns are fed into this station, followed by automatic welding into a frame that is used as the "chassis" or bearing structure during additional completion processes on the production line.

2. *Production flow system*: After automated welding, the steel frame chassis travels through the factory from workstation to workstation until it is fitted and finished with all installations. The Sekisui and Toyota factories have gates on both sides of the assembly lines to receive materials, parts, components, and prefabricated bath or kitchen modules required for the customized production of individual units, all of them JIT and JIS from cooperating suppliers or other company internal factories.

3. *Preinstallation*: The preinstallation of furniture and cables is an important part of the production strategy of Sekisui Heim and Toyota Home. The higher the degree of technical installation, the more efficient is the prefabrication process. The factory is the perfect environment for a fast and highly qualified installation of technical infrastructure (pipes, cabling, etc.), building technology and appliances since the units can be processed and finished three-dimensionally from multiple sides simultaneously.

4. *Zero-waste*: Throughout the process, in contrast with conventional construction, minimum waste is generated. Both Sekisui Heim and Toyota Home operate zero-waste factories. This can be achieved with the supply of modules fitting into the product structure without further processing or cut-off waste. Another step in reaching the target of zero wastage is the fastidious sorting of material waste for reuse and recycling. In contrast to conventional construction, the industrialized production of buildings in an SE is highly sophisticated in matters of resource circulation control.

Figure 5.15. View of the production line. A factory completes about 150 units (10 to 15 houses) per day. The maximum speed of production is one unit every 1.5 minutes. (Photo: Sekisui Heim)

Figure 5.16. At the last workstation within the factory, the finished units are prepared thoroughly for transport and handed over to a transport truck. (Photo: Sekisui Heim)

5. *Quality-oriented production*: In addition, through quality controls performed by robots and highly trained and qualified staff, quality is continuously and rigorously inspected after each production step – both by dedicated workers and by robots (see, for example, Figure 5.59). Every company has developed quality checklists with 200 to 300 different items for a single house to achieve early detection of mistakes and save time and cost. The end goal is to achieve an error-free product that further enhances the company's perceived reliability.

Rapid On-Site Deployment

After four to six weeks of factory-based manufacturing (including design finalization, work preparation and scheduling, material ordering and preparation, and final assembly of the units on the production line), the completed units are delivered to the on-site location where the building shall be erected JIT and JIS by the company's transport and logistics division (Figures 5.16 and 5.17). The finished steel frame units and prefabricated roof modules are usually assembled within one day. This means that the house becomes waterproof and wind-tight immediately and construction defects and quality losses (a major challenge in conventional construction, see **Volume 1, Section 4.1**) are therefore reduced to a minimum. Within the protected house, specially trained assembly workers (neither Sekisui Heim nor Toyota Home employs unskilled low-wage workers in their factories) complete the house and various remaining installations in less than one month. Following the assembly of the units, minor interior works and outside facilities are finished within a month. The following formula for the delivery process can be set up:

Figure 5.17. On-site assembly of the prefabricated units. (Photo: Sekisui Heim)

Total duration of production $= x + y + t_1 + t_2$

$x =$ preconsultancy/setup of contract (depending on the customer)
$y =$ configuration of the building (depending on the customer)
$t_1 =$ 4–6 weeks factory-based production including design finalization, work preparation and scheduling, material ordering and preparation, and final assembly of the units on the production line (fixed by company)
$t_2 =$ 4 weeks on-site assembly (fixed by company)

5.3.3 Factory Layouts and Process Design Strategies

Japanese prefabrication companies are famous for their production processes that allow complex high-level components such as panels or units to be produced on demand, in an OPF manner, and with a considerable degree of automation. A key element of these manufacturing processes is the factory layout design. Japanese prefabrication companies generally utilize automation in parts and low-level production/assembly, production planning and scheduling, factory internal material supply and micro-logistics (including material sorting and sequencing) to individual workstations. Final assembly on the main line, however, is done or guided by flexible, highly skilled human workers as central element heavily assisted by mechanical and robotic tools. The process and factory layouts are tailored to this strategy.

Factory Layout Design

Principles of advanced demand-oriented production, factory internal and external logistics, OEM models, supply chain design, flexibility and adaptability of manufacturing systems, sustainability in manufacturing, and future concepts in manufacturing were explained in **Volume 1, Section 4.3**. In the following section, the introduced knowledge will be completed by some specific knowledge about factory layouts and process designs necessary to understand the manufacturing processes used in the Japanese prefabrication industry.

Ultimately, factory layouts and process design strategies are core themes in technology and organization in manufacturing (see, e.g., Stephens & Meyers, 2013 and Thomopoulos, 2014). Factory layouts are the framework for the arrangement of the means of production (workstations, human workers, tools, storages, supply, logistics, material, etc.). They also determine processes and material flow patterns, and are especially important for the manufacturing of complex physical products (see **Volume 1, Chapter 5** for further examples of the manufacturing of complex products). Process design strategies can structure factory layouts on various levels, ranging from the micro level to the macro level (machine level, workstation level, group/cell level, segment level, factory level, network level, see also **Volume 1, Sections 4.2** and **4.3**).

The organizational view addresses the organizational relationships between individual stations or factory segments and the flow of material and information between them. Table 5.5 gives an overview of various manufacturing types classified according to the organizational view. Henry Ford's production system marked the transition from workbench-like organization to large-scale chain- or line-based manufacturing systems. However, this did not mean that workbench-like or workshop-like organization vanished. Rather, it meant that the dominant organizational principle changed and other organization methods were still used in parallel for subentities of production systems and sub-lines. Manufacturing systems, in general, are always combinations of the organizational forms outlined in Table 5.5. Today, as the examples of Dell, VW, Airbus (see **Volume 1, Chapter 4**) and Kajima (see **Volume 4**) show, workshop-like and workbench-like elements are predestined for being used to efficiently customize or personalize products and, therefore, become dominant elements in manufacturing once again. However, these organizational forms are now augmented by novel technology. As a result, the workbench-like organization, as in the case of Dell (utilizing augmented reality, intelligent guidance through workflows and in the near future probably smart glasses), for example, is far from a classical workshop-like approach. Nevertheless, in general, standardization and thus process stability in a workbench-like organization is lower than in a production-line organization.

In the Japanese prefabrication industry, two general types of organizational schemes are deployed:

1. *Production-line organization*: At unit manufacturers such as Sekisui Heim (steel units and wood units), Toyota Home (steel units), Pana Home (steel units), and Misawa Homes (Hybrid system, steel units), the main production unit is organized according to a production line organization. Sublines that deliver components to the mainline are, in general, organized according to flow-line organization, or also production-line organization, to be able to deliver the

Table 5.5. *Manufacturing systems analysed from an organizational point of view*

a. Workbench-like organization
The product remains at a fixed station in the factory, where it is produced or assembled manually or automatically with use of various tools. Means of production are organized around this one station. Synchronization with other stations or work processes is not necessary. Potentially, each product can be different.

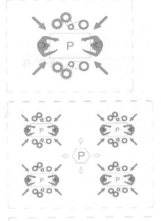

b. Workshop-like Organization
The product and/or its components flow between workstations. The sequence is not fixed nor is it fixed that the product must go through each station to be completed. Every workstation has a set of processes and tools bound to it.

c. Group-like organization
Individual workstations are bound together to groups. Those groups can refer to workstations with similar means of production or to workstations with means or production complementary to each other. Groups themselves can, for example, be organized as chains or flow line–like.

d. Flow line–like organization
Individual workstations do not have a fixed flow of material but a general direction of flow of material is common (e.g., within a factory segment of a factory).

e. Chain-like organization
The flow of material between individual workstations is highly organized and fixed. A material transport system linking the stations exists. However, the cycle times of the workstations are not synchronized and buffers between individual workstations may be used.

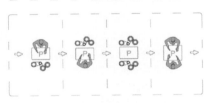

f. Production-line organization
The flow of material between individual workstations is fully determined and organized. A material transport system linking the stations exists. The cycle times of the workstations are fully synchronized and no buffers between individual workstations are needed.

g. Explanation of symbols

Tool/machine

Workflow

Parts delivery

Product

Drawings by B. Georgescu.

components in full synchronization with the time-phased assembly on the main line.

2. *Chain-like organization combined with elements of flow-line organization*: All non-unit manufacturers, in general, apply organizational schemes that can be situated between the chain-like organization (general factory organization) and flow-line organization (e.g., in the manufacturing of frames). In some parts of the factory, production line–like organization can also be found, although it is not the dominant element as in unit manufacturing.

Suppliers of course apply a variety of organizations ranging from simple workbench-like organization (window suppliers) to production-line organization (e.g., bath module suppliers).

Combination of Automation and Human Flexibility

In **Volume 1, Chapter 5**, it was shown that buildings, similar to ships and aircrafts due to their size, weight, and complexity, are products that in many cases cannot be produced on production line–like settings. Japanese prefabrication companies circumvent this problem by changing the modular structure of buildings and splitting them down into a set of units or panels that can then be manufactured in line-based settings in SEs in the factory. Nevertheless, these units and panels are still complex entities and, as discussed earlier in this chapter, each unit or panel is customized/personalized and thus different. Japanese prefabrication companies tackle the task of automating the highly complex final assembly of these customized/personalized units or panels in the factory by automating in particular parts and low-level production/assembly, production planning and scheduling, factory internal material supply and micro-logistics (including material sorting and sequencing) to individual workstations. The actual assembly on the main assembly line, however, is done or guided by flexible, highly skilled and technology augmented human workers. These human workers are usually assisted by specialized tools, mechanical equipment, and robotic equipment (e.g., balancers and collaborative robots, automated logistics/micro-logistics systems, automated guidance through work flows). All in all, it can be said that the Japanese strategy to deal with the integration of automation ratio and human flexibility at the final production lien where high flexibility is required is based on the idea of "automating in the background". Therefore, the settings around the final assembly line are optimized for both automated supply and efficient technology augmented human workers.

Prefabrication based on three-dimensional steel frames and strategies to integrate automation and human flexibility in Japanese prefabrication which has existed in its present form since the 1980s has similar characteristics to VW's Phaeton production strategy. It is speculation whether VW was inspired by the Japanese prefabrication companies or not. However, it is clear that VW faced a similar problem as, for example, Sekisui Heim: how to manufacture a complex product with a batch size of one with a certain degree of automation on a production line utilizing the skill of highly trained workers.

In 2002, Volkswagen opened a factory in Dresden (Gläserne Manufactur; Figure 5.18) for the production of the VW Phaeton. The VW Phaeton is a luxury car with

a) Final assembly line: final assembly is performed by a flexible human worker but in the background logistics robots provide all materials just in time

b) Racks with parts and equipment are brought by logistics robots to the worker.

Figure 5.18. Flexible production of the Volkswagen Phaeton in the Gläserne Manufaktur in Dresden, Germany. (Photos: VOLKSWAGEN AG)

c) Special jigs and fixtures orientate the car bodies to simplify assembly operations by the human workers.

Figure 5.18 (*continued*)

a high degree of customization/personalization. The configuration of each car on the conveyor belt varies more than in any other car series of Volkswagen. Therefore, Volkswagen uses no robots for positioning or assembly on the production line but fully exploits the flexibility of the human workers. VW marks its Phaeton as a handmade car. With a price of about €80,000 to 100,000, the Phaeton is relatively inexpensive for a handmade, customized/personalized car. VW achieves this efficiency through "hidden" automation. Although all assembly processes at the main line are manual, the logistics/transport of parts and components to the production line (where the car bodies and the assembly workers are) is fully automated. The racks between cars (Figure 5.18a and b) contain all necessary parts and are continuously exchanged by robotic, automated guided vehicles (AGVs) that can couple into the racks. This exchange takes place JIT. Once a rack is close to empty, another one with parts for the next process is already waiting beside the conveyor to replace the empty one. Furthermore, robotic jigs and fixtures more and orientate the car bodies in order to simplify assembly operations by the human workers.

VW is utilizing the flexibility of the human worker for allowing for the flexibility of assembling sequences necessary to manufacture customized/personalized products in an OPF-manner. The human flexibility is supported by a high-tech, cooperative equipment and highly automated (cellular) logistics systems (Figure 5.18c). Nevertheless the Phaeton assembly has similarities with the final assembly of

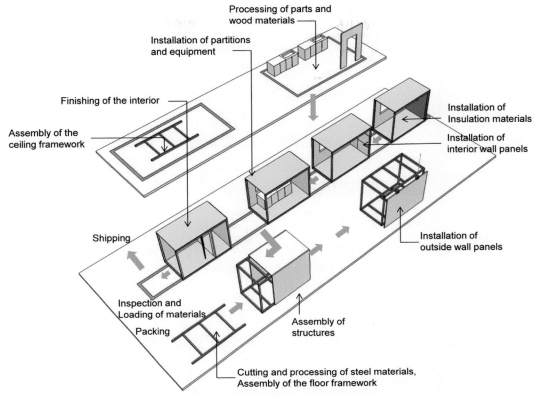

Figure 5.19. General idea of Sekisui Heim's production line principle. (Drawing on basis of Sekisui Heim)

three-dimensional building modules (each one is different here as well) in the factories of Sekisui Heim, the assembly and customization of the Phaeton in some aspects can be considered as more complex (e.g., in terms of accuracy and quality demanded and the type of technologies and sub-systems installed) than the assembly of a customized unit for a building on the production line. For this reason, the combination of human flexibility and automated parts delivery in the Gläserne Manufaktur (see also VW, 2013) represents the next step in production line evolution from which also the Japanese manufacturers might learn in the future.

Sekisui Heim – General Idea of Factory Layout

Similar as, for example, Toyota Motors with its flexible body line, Sekisui Heim first automatically generates a three-dimensional space frame that serves as a component carrier during the production line–based final assembly. In the case of Sekisui Heim, the whole manufacturing layout and process design are strictly based on the production line principle (Figures 5.19 to 5.22). In contrast, Figures 5.24 to 5.28 show that other manufacturers (using other materials, modularization principles, and manufacturing strategies) do more apply chain-like organization types combined with elements of flow-line organization.

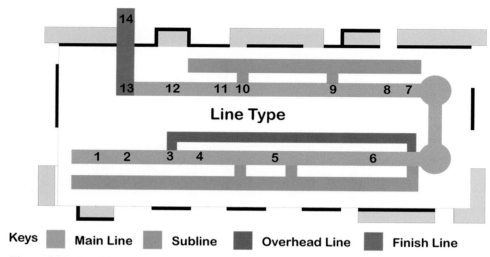

Figure 5.20. Combination of sub-line types in a standard layout of a Sekisui Heim factory. (Drawing by W. Pan)

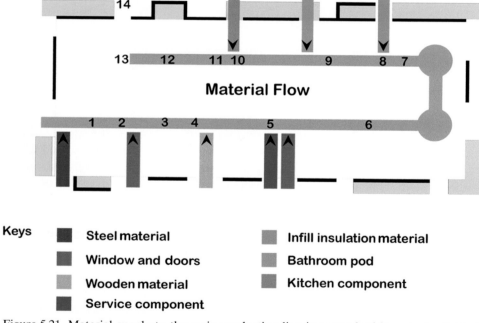

Figure 5.21. Material supply to the main production line in a standard layout of a Sekisui Heim factory. 1 = steel material cutting and drilling station; 2 = welding station; 3 = frame fabrication/nailing; 4 = ceiling panel/floor board installation; 5 = automatic welding; 6 = side wall installation; 7 = insulation installation, 8 = internal insulation installation, 9 = bath components/modules installation; 10 = kitchen installation; 11 = final finishing; 12 = final inspection; 13 = packing; 14 = ready for delivery. (Drawing by W. Pan)

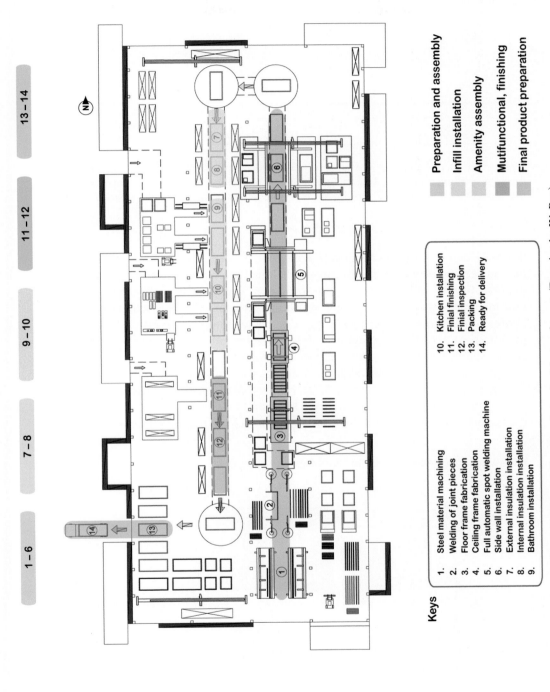

1 – 6   7 – 8   9 – 10   11 – 12   13 – 14

Keys

1. Steel material machining
2. Welding of joint pieces
3. Floor frame fabrication
4. Ceiling frame fabrication
5. Full automatic spot welding machine
6. Side wall installation
7. External insulation installation
8. Internal insulation installation
9. Bathroom installation

10. Kitchen installation
11. Finial finishing
12. Finial inspection
13. Packing
14. Ready for delivery

■ Preparation and assembly
■ Infill installation
■ Amenity assembly
■ Mutifunctional, finishing
■ Final product preparation

Figure 5.22. Sekisui Heim's production line-based standard factory layout. (Drawing by W. Pan)

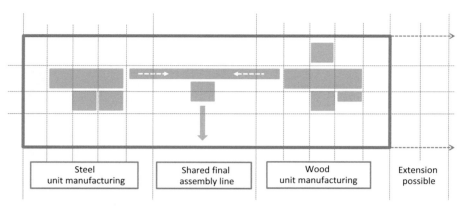

| Steel unit manufacturing | Shared final assembly line | Wood unit manufacturing | Extension possible |

Figure 5.23. In the modular and flexible Hokkaido factory steel and wood unit manufacturing is combined into one factory with a shared final assembly line. The centre, and thus the resources of the factory, can be shifted depending on the demand for wood or steel construction.

Sekisui Heim – Hokkaido Modular Factory

For more than 30 years, the Japanese prefabrication companies have developed modular factory and resource supply structures. The recurring introduction of new housing types brings along changes in product structures and consequently, applications, tools, and machinery must be added, exchanged, or adjusted dynamically over time. Materials and component delivery organization must also be modular to enable supply chains to react to changes in demand.

An interesting example for an up-to-date modular and flexible factory is Sekisui's new factory near Sapporo (Figure 5.23). In this factory, steel frame and wood frame production is combined in one factory. The steel frame units stream from one side of the factory to the centre, and the wood frame units from the other side to the centre.

Misawa Homes – Wood Panel Manufacturing Layout

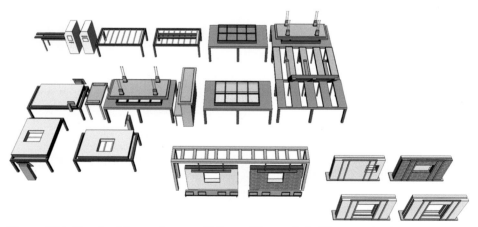

Figure 5.24. Standard factory layout of Misawa Home: chain-like organization type combined with elements of flow-line organization (simplified representation). (Drawing by W. Pan)

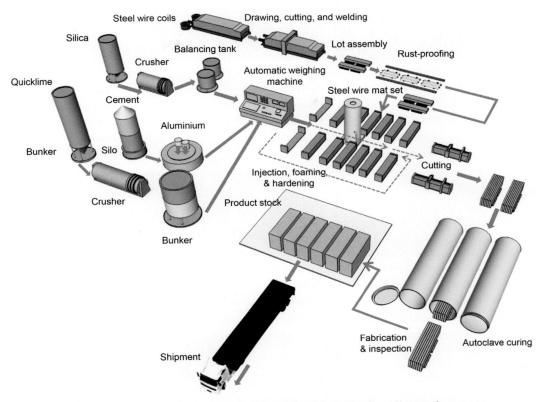

Figure 5.25. Standard layout Asahi Kasei/Hebel House for the production of aerated concrete elements: chain-like organization type combined with elements of flow-line organization. (Authors' representation of information given by Asahi, 2014) (Drawing by W. Pan)

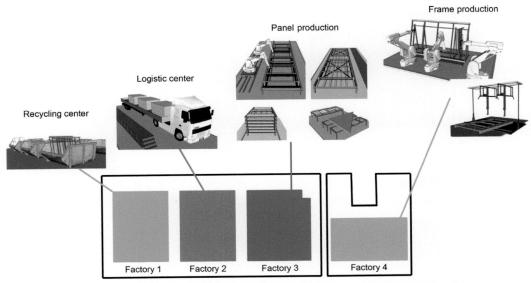

Figure 5.26. Standard factory layout of Daiwa House: chain-like organization type combined with elements of flow-line organization (simplified representation).

a) 1. Tohoku factory

Products: Single-family,
housing complexes,
business/industry
building construction

Location: Osaki, Miyagi

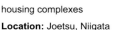

b) 2. Niigata factory

Products: Single-family houses,
housing complexes

Location: Joetsu, Niigata

c) 3. Tochigi Ninomiya factory

Products: Single-family houses,
housing complexes,
business/industry
building construction

Location: Mooka, Tochigi

d) 4. Ryugasaki factory

Products: Single-family houses,
housing complexes

Location: Ryugasaki, Ibaraki

e) 5. Chubu factory

Products: Business/
industry building construction

Location: Fukuroi, Shizuoka

f) 6. Nara factory

Products: Single-family houses,
housing complexes

Location: Nara

g) 7. Mie factory

Products: Single-family houses,
housing complexes

Location: Komono, Mie

h) 8. Sakai factory

Products: Business/
industry building construction

Location: Sakai, Osaka

i) 9. Okayama factory

Products: Single-family houses,
housing complexes

Location: Akaiwa, Okayama

j) 10. Kyushu factory

Products: Single-family houses,
housing complexes,
business/ industry building construction

Location: Kurate

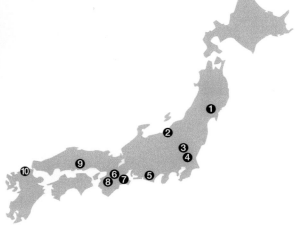

k) Factory locations throughout Japan

Figure 5.27. Overview factory network of Daiwa House. (Photos: Copyright Daiwa House Industry Co., Ltd. All rights reserved.)

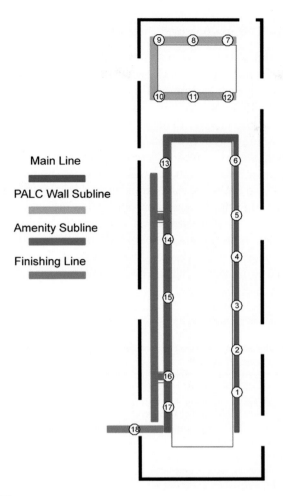

Main Line

PALC Wall Subline

Amenity Subline

Finishing Line

Factory Plan:

1) Steel machining; 2) Welding; 3) Floor panel fabrication; 4) Ceiling panel fabrication;5) Capsule unit spot welding; 6) Wall installation; 7) Re-bar mesh fabrication; 8) Rust-proofing for mesh; 9) Ceramic concrete pouring; 10) Moulding; 11) Autoclaving; 12) Painting; 13) Kitchen unit installation; 14) Bathroom unit installation; 15) Final finishing; 16) Final inspection;17) Packing; 18) Delivery

Factory Sections:

Preparation: 1-6; PALC wall: 7-12; Amenity installation: 13-14; Final assembly: 15-18

Figure 5.28. Standard factory layout Misawa (Ceramic Home Hybrid Model Series): combination of production line with flow-line organization (simplified representation).

Finishing is done in the centre area for both wood and steel frame units. The centre, and thus the resources of the factory, can be shifted depending on the demand for wood or steel construction. This modular and flexible factory allows the company to respond to future changes in wood or steel construction demand by reorganizing and adapting internal organization.

5.4 Analysis of Selected Companies and their Manufacturing Systems

In the following section, successful Japanese companies and their diverse backgrounds, building systems, manufacturing strategies, and business concepts are outlined. To unify the outline of each system and make them comparable, a set of thematic fields and related influencing factors was defined serving as a common reference ground throughout this chapter (Table 5.6).

Furthermore, each building system (Table 5.7) is described graphically and pictures of the manufacturing process, factories components, and buildings create, together with the other given information, a holistic view of the companies and their systems. The chapter also identifies the unique characteristics of the different strategies and approaches within the Japanese prefabrication industry. The holistic view on each company reveals that a substantial number of the most advanced prefabrication companies are typically a subsidiary of a larger company from the chemical and material engineering industries (e.g., Sekisui, Asahi Kasei) or from the electronics, automation or automotive industries (Panasonic, Sanyo, Toyota). These

Table 5.6. *Thematic fields and related influencing factors serving as a common reference ground throughout this chapter*

Thematic field	Influencing factors
History	Year of foundation
	Background, origin, and information about mother company
	Business fields of the company
	Timeline of development of the company
Business volume	Turnover per year
	Output per year
	Increasing/decreasing sales
	Highest sales ever reached
Employees and factories	Number of factories
	Distribution of factories
	Number of employees
	Salaries, ratio of engineers vs. ratio of assembly workers
Range of products	What types of prefabricated buildings are offered?
	What other types of products are offered?
Business strategy	Which market segment is targeted?
	Quality of products
	Synergies created with other company divisions and their products and services
	Quality of customer service
Manufacturing strategy	Manufacturing layout of factory
	Material flow in the factory
	Degree of prefabrication and automation ratio
Other Specialties	Innovative approaches
	Cooperation with other companies
	Outstanding research and development activities
	Excellent technology (e.g., earthquake resistance performance, sustainability, etc.)
	Subsidiaries outside of Japan

Table 5.7. *Overview of companies outlined in this chapter*

	Company/system	System type
1	Sekisui House	Fully panelized steel kit
2	Daiwa House	Steel frame combined with steel panels
3	Pana Home	Steel panels combined with individual steel elements
4	Sanyo Homes	Steel frame combined with steel panels
5	Asahi Kasei – Hebel House Homes	Steel frame combined with aerated concrete panels
6	Misawa Homes – sub- and mini-assembly units	Wood panels
7	Mitsui	Wood panels
8	Tama Home	Wooden frame combined with panels
9	Muji House	Wooden frame combined with panels
10	Sekisui Heim	Steel units
11	Toyota Home	Steel units
12	Misawa Homes – Hybrid	Steel units
13	Sekisui Heim – Two-U Home	Wood units

subsidiaries create additional deamand for the products of the mother company, serve as a platform for sales and awareness of multiple products and businesses of the parent company, and are in most cases the key customers of the products of the parent company. Those parent companies, therefore, not only support the prefab housing businesses financially in building up advanced manufacturing systems and factories, but also provide them with necessary knowledge and know-how.

5.4.1 Sekisui House (Fully Panelized Steel Kit)

Product structure	Building is comprised of two-dimensional steel-based and preassembled panels (Figure 5.30)
Manufacturing layout	Flow line–like and group-like organization. Focal point of the production/assembly is the station where robots assemble and weld the basic steel frame structure. High depth of added value (even steel profile pressing is done on factory site)
Material flow	Lean but not fully continuous; use of buffers and storage places, automated warehouse
On-site assembly	More than 40% of work remains on the construction site, assembly on-site takes about 2–3 months

History

Sekisui House originated from Sekisui Chemical Corporation and was founded in 1960. From the beginning, it was involved in activities that promoted and advanced the transition to steel-based prefabricated buildings in Japan. In contrast to Sekisui Heim, which still belongs to Sekisui Chemical Corporation, serving as its housing division, today Sekisui House is an independent corporation. Sekisui House has consistently been among the top sellers in the industry and has up sold about 2 million prefabricated houses to date. For more detailed information, see Sekisui House (2015).

Figure 5.29. Automated steel frame assembly and welding station. (Photo: Sekisui House)

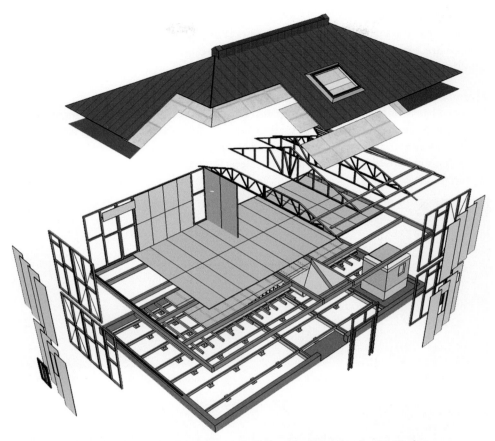

Figure 5.30. Explanation of Sekisui House's building kit. (Drawing by W. Pan)

Business Volume
Sekisui House has a yearly turnover of about €15 billion and currently produces about 50,000 buildings per year. The highest amount of buildings sold in a year was about 78,000 units in 1994.

Employees and Factories
Sekisui House currently has about 15,000 employees and operates five main factories. All factories are certified according to the ISO 14001 standard on advanced environmental management. Sekisui House's average annually salary per employee is about €70,000 after taxes and thus is ranked at the top in the industry.

Range of Products
Sekisui House focusses on fully panelized steel kits and, because of the growing importance of wood in Japan, also offers wooden frame kits. In 2012, it installed a new and advanced production line for panels that can be attached to its wooden frame kits in its Shizuoka factory.

Business Strategy

The company focusses on the upper market segment and, in particular, its steel houses are regarded as (aside from Daiwa House and Sekisui Heim products) the qualitatively most advanced houses available in Japan.

Manufacturing Strategy

Sekisui House has set up a chain-like and flow-line–like organization in most factories. Central elements of the production line are the robotic assembly and robotic welding station that are directly combined and produce the basic panel frames (which serve then in subsequent manufacturing steps as chassis/template) completely automated (Figures 5.29 and 5.31).

a) Production of steel profiles

b) Automated assembly of steel frames

c) Automated welding of steel profiles on the production line

Figure 5.31. Outline of key steps in Sekisui House's manufacturing process. (Photos: Sekisui House)

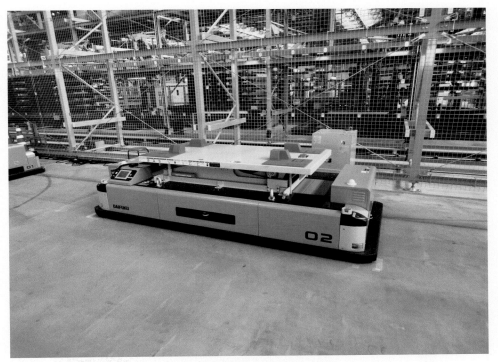

d) Factory internal logistics system

e) View of the production line

Figure 5.31 (*continued*)

f) Fully automatic welding

g) Automated coating of steel frames

Figure 5.31 (*continued*)

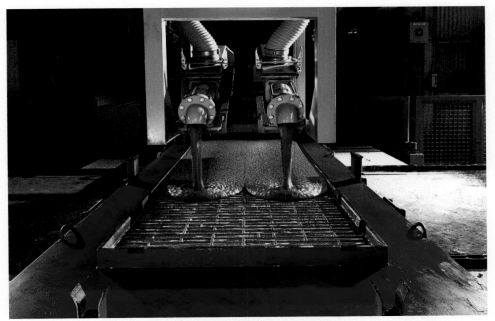

h) Automated production of exterior wall

i) Automated coating of exterior wall

Figure 5.31 (*continued*)

Other Specialties

Sekisui House is one of the largest home builders in Japan. Besides, Sekisui House is a member of the Prefab Club (Japan Prefabricated Construction Suppliers and Manufacturers Association) and thus a major supplier of buildings following natural disasters. Sekisui House has also built up business divisions in Australia and in China.

j) Computer-controlled hardening of finishing

k) Automated transfer and quality control of exterior wall components (in this case, exterior wall components for timber houses)

Figure 5.31 (*continued*)

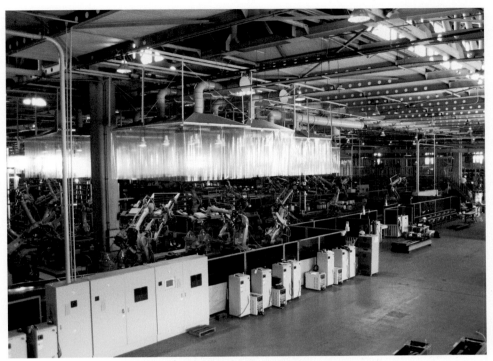

l) View of the steel frame production section in a factory

m) Steel frame assembly and welding by a multi-move robot operation
(=simultaneous activity/movement of robots to process a work piece)

Figure 5.31 (*continued*)

n) Material recycling station

o) Example of final product

Figure 5.31 (*continued*)

5.4.2 Daiwa House (Steel Frame Combined with Panels)

Product structure	Building is composed of two-dimensional steel-based and preassembled frames (Figure 5.33).
Manufacturing layout	Flow-line–like and group-like organization. Systemized work environment and work processes but high ratio of human labour (e.g., most welding and logistics tasks are performed by human labour).
Material flow	Not continuous; many buffers and storage places.
On-site assembly	More than 40% of work remains on the construction site, assembly on-site takes 1–2 months. Advanced labour management concerning on-site assembly: Daiwa House cooperates with and efficiently trains local builders and subcontractors.

History

Daiwa House Industry Corporation was started in 1955, about five years earlier than Sekisui House. In the immediate postwar period, Daiwa House supplied urgently needed shelters on the basis of a steel pipe structure onto which a variety of panels could be mounted. Inspired by the promotion of prefabricated steel buildings at that time, Daiwa House launched its first steel frame– and panels–based house in 1959. Since its founding, Daiwa House has consistently been among the top sellers in the industry. Today it is one of the fiercest competitors of Sekisui House in the fight for the top position. For more detailed information, see Daiwa House (2015).

Business Volume

Daiwa House currently has a yearly turnover of about €11 billion and supplies about 43,000 buildings each year. Daiwa House reached its highest annual sales in 2007, when it sold nearly 45,000 units. Considering Daiwa House consistently strengthens its developer approach and has recently acquired the contractor Fujita, it can be assumed that its sales volume will increase significantly in the next years.

Employees and Factories

Daiwa House operates about 10 main factories and employs about 13,500 people (see also Figure 5.27). Its average annual employee salary is about €75,000, also among the highest in the industry. In contrast to Sekisui House and other manufacturers of prefabricated steel-based buildings, Daiwa House extensively subcontracts on-site assembly work to local builders and craftsmen. Quality standards in its factories are outstanding, and already in 1974, Daiwa House received an award from the Japanese government for its quality management method.

Range of Products

Daiwa House focusses on fully panelized steel kits but also offers wooden buildings. Besides single-family houses, Daiwa House has a strong foothold in the condominium business, where it also applies its kit strategy.

Business Strategy

Daiwa House serves the mid to top end of the market. In the top end market, Daiwa House probably delivers the most customized buildings in the industry. Furthermore,

Figure 5.32. Automated steel panel production. Robots weld the profiles together, which are placed on a jig/template. (Photo: Copyright Daiwa House Industry Co., Ltd. All rights reserved.)

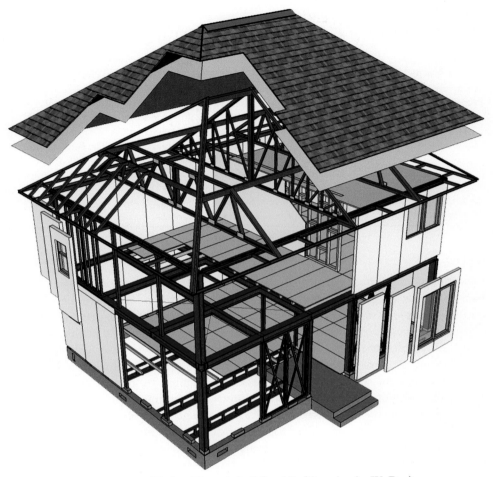

Figure 5.33. Explanation of Daiwa House's building kit. (Drawing by W. Pan)

Daiwa House not only sells buildings to end customers in the conventional approach, but also appears as a developer that sells and rents its own developments, and therefore, in a way, "buys" its own prefabricated buildings.

Manufacturing Strategy

Daiwa House has established flow-line–like and group-like organization (see also Figures 5.26 and 5.27). It provides a SE and utilizes robots for key assembly tasks (Figure 5.32), although its automation ratios in some factories are slightly lower than those of Sekisui House and Sekisui Heim.

Other Specialties

Already in 1989 Daiwa House established its "Silver Age Research Centre", and today, with the worldwide issue of age-related demographic change, is one of the leading home builders in terms of universal and age supportive designs and building features. Today, Daiwa House cooperates with leading technology companies to integrate advanced assistance technologies into its buildings (e.g., intelligent toilet developed together with Toto). Daiwa House is currently trying to strengthen its business in China.

5.4.3 Pana Home (Steel Panels Combined with Steel Components)

Product structure	Buildings are modularized into large-sized steel panels and steel components (columns, beams, etc.) that are prefabricated in the factory (Figure 5.35).
Manufacturing layout	Factories are organized in a flow-line–like manner. The steel panels are placed in specialized steel fixtures that present the panels in a convenient way to the workers and guide the worker and machines during the assembly process.
Material flow	Lean but not fully continuous; use of buffers and storage places.
On-site assembly	Assembly on-site takes about 2–3 months.

History

Panasonic is a multinational electronics company that has continuously expanded its business fields since it was founded. Panasonic's expansion has always been based on the principle of transferring its electronics know-how into the new field and to expand the market of its core products. Just as Microsoft recently acquired Nokia to be able to produce better software–hardware integrated communication devices in the future, Panasonic set up its housing division to be able to better integrate built environments with their diverse set of electronics. In light of upcoming key issues such as energy efficiency and ubiquitous computing, Panasonic's move into the building industry has started to really pay off and Pana Home, during the last decade alone, became one of the top five prefabricated housing sellers in Japan. Accordingly, Pana Home, which integrates Panasonic products into its buildings wherever possible, became one of the biggest "customers" for Panasonic's electronic products.

Panasonic was founded 1918 under the name Matsuhita Electric. In particular, after the Second World War, a worldwide expansion began. Panasonic was a brand name of Matsuhita and in 2008, all subsidiaries including the Japanese mother company were changed in a unified way towards Panasonic as a brand and company name. In the 1950s, Matsuhita's building material division started to develop prefabricated buildings and in 1977, the division adopted the brand name Pana Home. For more detailed information, see Panasonic (2015).

Business Volume

Pana Home sales have been on a steady increase, particularly in the last decade. Pana Home has a yearly turnover of about €2.6 billion. It currently sells about 11,000 prefabricated houses annually. Panasonic's activity in the field of green electronics and sustainable cities is likely to further increase its housing sales in the future.

Employees and Factories

Pana Home operates two main factories and employs about 5000 people.

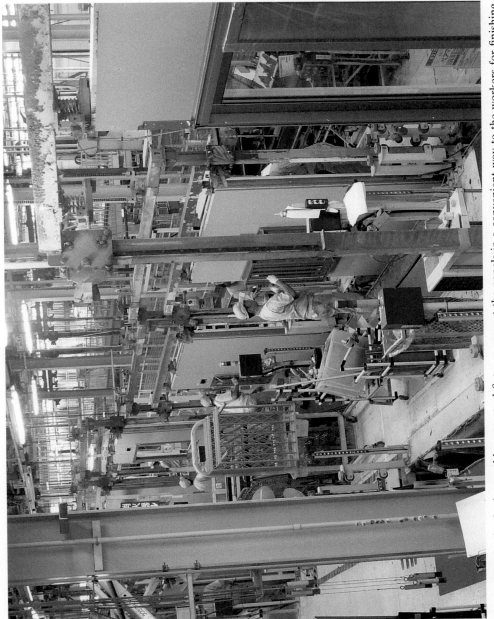

Figure 5.34. In the final assembly area steel fixtures present the panels in a convenient way to the workers for finishing. (Photo: Pana Home)

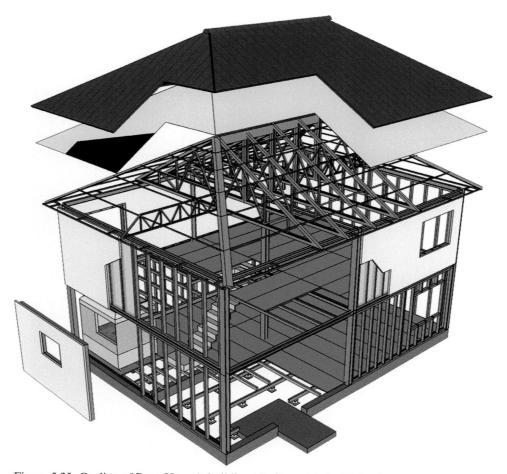

Figure 5.35. Outline of Pana Home's building kit. (Drawing by W. Pan)

Range of Products

Pana Home focusses on steel-based panelized houses. For its Smart Eco House
series, using wood as building material is under consideration in light of the sustainability discussions in Japan and growing demand for wood-based buildings. The
intended wood versions are very similar to the steel versions and are modularized
and panelized in nearly the same way.

Business Strategy

Pana Home focusses on the upper market segments, and in particular, on customers who want houses with smart home features but at a reasonable price. Products
are designed to incorporate the latest Panasonic electronics products, lighting systems, and home appliances. In particular, Pana Home buildings incorporate smart
home technology almost as standard. For example, its smart home energy management System SMARTHEMS™ (Smart Home Energy Management System) and
its system AiSEG™ (Artificial Intelligence Smart Energy Gateway), which connect
the electrical equipment and appliances around the home for smart energy saving,

a) Step 1

b) Step 2

c) Step 3

d) Step 4

e) Step 5

f) Step 6

g) Step 7

Figure 5.36. Pana Home's on-site assembly process. (Photos: Pana Home)

are often implemented. Many models of Pana Home also integrate solar modules supplied by Panasonics solar panel production business as standard equipment. Furthermore, customers have the option of installing Panasonics fuel cell Ene Farm. With Panasonic's plans to become the world's leading "green electronics" manufacturer, Pana Home becomes more and more important for Panasonic as a customer and promoter of such technologies and products. The built environment in general, and in particular buildings, consume a tremendous amount of resources and energy, and a close integration between buildings and electronics performance as practiced by Panasonic and its housing division is considered by science and industry more and more key for tackling this problem.

Manufacturing Strategy
Pana Home factories are organized in a flow-line–like manner. The steel panels are placed in specialized steel fixtures that present the panels in a convenient way to the workers and guide the worker and machines during the assembly process (Figure 5.34). Figure 5.36 outlines the remaining on-site assembly process steps.

Other Specialties
Aside from the incorporation of advanced smart home technology for a reasonable price, Pana Home buildings are famous for their advanced security systems. Here, Pana Home can directly use Panasonic's knowledge and experience in security and surveillance systems. Furthermore, Pana Home has recently built up subsidiaries in Taiwan, Malaysia, and Vietnam.

5.4.4 Sanyo Homes Corporation (Steel Frame Combined with Panels)

Product structure	Light steel framework structure (Figure 5.37): • Steel frame: pillar, beam • Roof frame: steel frame • Floor frame in the first floor: steel frame with wooden panel • Other floor frames, external wall, and roof: wooden panels
Manufacturing layout	Factory is divided into different segments in which the different parts for its building kit are produced.
Material flow	Flow-line–like material flow.
On-site assembly	Off-site production takes about 1 month and assembly on-site about 2–3 months.

History
Kubota House was founded in 1969 by Kubota Corporation. In 1987, Sanyo Electric Corporation established its Sanyo Real Estate division and entered the housing and condominiums industry in the 1990s. In 2002, Sanyo Electric Corporation fused its real estate and construction business with Kubota House and gave the conglomerate the name Sanyo Homes Corporation. Sanyo Homes was at that time fully owned by Sanyo Electric Corporation. Between 2002 and 2014 changed its ownership structure radically and the LIXIL Group, Orix, Kansai Electric Power and Secom are now the main shareholders. For more detailed information, see Sanyo Homes (2015).

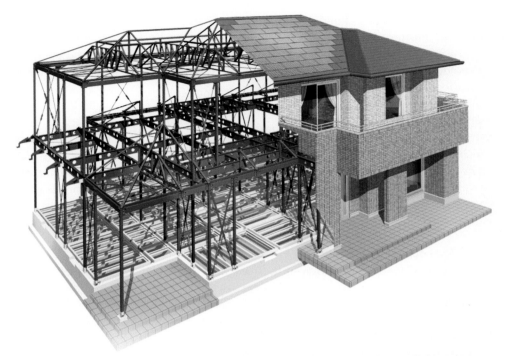

Figure 5.37. Design example and explanation of the building kit of Sanyo Homes. (Photo: Sanyo Homes Corporation)

Business Volume
Sanyo Homes has a yearly turnover of about €0.3 billion and sells about 1000 buildings per year. For 2014 the company is expecting an output considerably higher than 1000 buildings.

Employees and Factories
Sanyo Homes operates one main factory located in the Osaka area and has about 700 employees.

Range of Products
Sanyo Homes before 2002 focussed solely on steel-based panelized buildings and combined a bearing steel frame kit with steel panels. Similar to other manufacturers, since 2002, Sanyo Homes intensified activity in developing and selling electronic-based smart home products and in developing areas such as Smart Towns that can then be built up with its own buildings (e.g., Suma-e-Town in Tsurumi-ku, Osaka). From 2006 on Sanyo broadened its field of activity even more and is now offering housing and condominium products, smart homes and smart home electronics installation, and various services related to remodeling, life support, and asset utilization.

Business Strategy
Sanyo Homes produces high-quality buildings but does not focus on the high-end market. Most of the building sold can be considered to be in the middle-range market

segment. Sanyo Home's buildings are, of course, also vehicles for selling products from other business divisions of its shareholders (Sanyo, LIXIL Group, Orix, Kansai Electric Power, Secom) such as home appliances, electronics, smart home/building technology components, and services related to remodeling, life support, and asset utilization.

Manufacturing Strategy

Sanyo Homes' main factory in the Osaka area is divided into different segments in which the different parts for its building kit are produced (e.g., floor panel line and outer wall panel line). These sections share cross-sectional installations for painting or welding. Its research and testing facilities are located at the Osaka factory as well, and ensure a closed loop feedback in terms of quality standards, improvements, and innovations.

Other Specialties

Sanyo Homes' name recognition in the past was not as high as that of other top prefabrication companies, as it didn't advertise very much. However, since 2006 and with new shareholders and with the increase of its activity filed (smart homes, e-towns, services, etc.) Sanyo Home continuously strengthened its recognition as a provider of high-tech products and services. Sanyo Homes is also well known for its high standards concerning safety and security technology built into its buildings. Furthermore, in 2014, Sanyo Homes opened its Sanyo Homes Carpenter School.

5.4.5 Asahi Kasei – Hebel House Homes (Steel Frame Combined with Aerated Concrete Panels)

Product structure	Bearing steel frame structure consisting of prefabricated components and panels. Aerated and insulated concrete panels are mounted to this steel structure (Figure 5.38).
Manufacturing layout	In general flow line–like. In some areas, as, for example, in the production of the aerated concrete panels, stricter chain-like organization is deployed.
Material flow	Lean but not fully continuous; use of buffers and storage places.
On-site assembly	Assembly on-site takes about 2–3 months.

History

Asahi Kasei is a multinational company that has established, around its core competencies in chemicals and material engineering, a multitude of business fields in which it is active with its divisions or subcompanies. It began operation in 1922 as a chemicals producer, and then moved step by step into the materials engineering business and the petrochemicals industry, and in 1957, it introduced its plastics business. In 1967, Asahi Kasei entered the construction material business with its autoclave aerated concrete panels (Hebel panels). In 1972, Asahi Kasei established its housing business and introduced its steel-frame–based system that is equipped with Hebel panels serving as exterior and interior walls. In 1982, Asahi Kasei started its

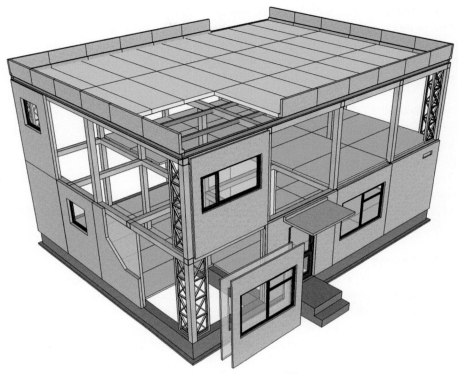

a) Abstract representation of the building system

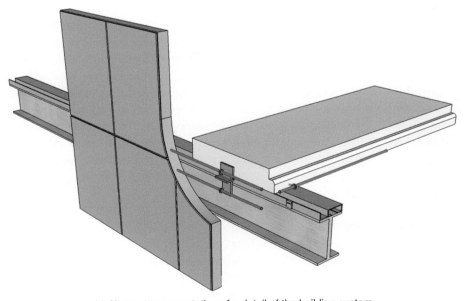

b) Abstract representation of a detail of the building system

Figure 5.38. Explanation of Asahi's building kit. (Drawing by W. Pan)

semiconductor and electronics business. Today, Asahi Kasei is active in a multitude of business fields ranging from chemicals, fibres, homes, and construction materials to electronics and healthcare. Closely related to Asahi Kasei is Asahi Glass, one of the biggest glass and glass panel manufacturers in Japan. For more detailed information, see Asahi Kasei (2015).

Business Volume

Asahi Kasei's housing division has a yearly turnover of about €4.5 billion. It currently sells about 16,000 prefabricated houses per year. The highest output ever reached was 18,195 buildings in 2012. The output per year is steadily rising.

Employees and Factories

Asahi Kasei's housing division operates two major factories and employs about 5000 people.

Range of Products

Asahi Kasei's housing division focusses on prefabricated buildings (houses and apartment buildings) made of its steel frames and aerated concrete panels. It is also active in real estate, remodelling, mortgage financing, and asset management services. Furthermore, aside from selling totally prefabricated buildings, it also sells subcomponents of its housing kit (e.g., standard and custom-made aerated concrete panels, thermal insulation, structural components, etc.) as individual products to builders and contractors.

Business Strategy

The specialty of Asahi Kasei's prefabricated buildings is the combination of a steel structure (steel panels and components) with elements of massive construction (aerated concrete). Asahi Kasei has adopted this technology for efficiently making such relatively light, fireproof, and insulating aerated concrete panels from the German company Hebel Haus. In Germany, massive construction is the most popular construction method for housing, and aerated concrete provides a modern alternative to conventional massive concrete or brickwork construction. Asahi Kasei reinforces aerated concrete and has developed connectors to its steel frame through which, despite the massive appearance, earthquake resistance is achieved. Asahi Kasei addresses the middle to upper market segment but only a small portion of the very top end market segment.

Manufacturing Strategy

The production process is organized generally in a flow-line–like manner (see also Figure 5.25). In some areas, for example, in the production of the aerated concrete panels, stricter chain-like organization is deployed. The high degree of automation of the reinforced aerated concrete panels is particularly interesting. Here, nearly the whole process, from material weighting to reinforcement meshes production to moulding and autoclave curing, is automated, ensuring the high quality of the panels.

Other Specialties

Through the combination of steel frames and aerated reinforced concrete, Asahi Kasei's buildings are highly earthquake resistant. Furthermore, in contrast to prefabrication-type buildings based solely on steel or wood as the main building material, Asahi Kasei's buildings still provide advantages of classic massive construction, for example, slow heating up and cooling off, fire resistance, resistance to humidity, and natural insulation capability. Asahi Kasei is considered as producing the most resistant buildings and components in terms of resistance to fire, wind, and rain.

5.4.6 Misawa Homes Sub- and Mini-Assembly Units (Wood Panels)

Product structure	Buildings composed of small or large wooden panels
Manufacturing layout	Misawa's factories are organized flow line–like with some sections as, for example, final assembly lines being based on production lines
Material flow	Lean but not fully continuous; use of buffers and storage places
On-site assembly	On-site assembly of wooden panel structures takes 2–4 months

Figure 5.39. Finishing of wood panels in a factory of Misawa. (Photo: Misawa Homes)

Figure 5.40. Automated handling of a large wood panel. (Photo: Misawa Homes)

History

Misawa began operation in 1962. Initially, Misawa focused mainly on the prefabrication of wood-based buildings. Today, wood-based buildings (sub- and mini-assembly units) are still the main focus of the company but it has added a model series based on steel units filled with wood components (Ceramic Home Hybrid model series, see also Section 5.4.12). For more detailed information, see Misawa (2015).

Business Volume

Misawa currently has a yearly turnover of about €3.7 billion and supplies about 12,500 buildings a year.

Figure 5.41. Assembly of finishing to the wood panels. (Photo: Misawa Homes)

Employees and Factories
Daiwa House operates about 12 main factories and employs about 9000 people.

Range of Products
Misawa traditionally has had a major focus on wooden building prefabricated components, prefabricated small panels, prefabricated large panel structures, and also later gained a foothold in the steel-based housing business. It is therefore one of the prefabrication companies with the broadest ranges in terms of housing and housing construction types.

Business Strategy
Accordingly, with its product range, Misawa's buildings cover a broad range of quality and cost levels from low-end to high-end market segments. Misawa's business strategy is to market this diversity and, at the same time, build up this diversity on a very standardized set of core competences, elements, and machinery.

Manufacturing Strategy
In general, in wood building manufacturing, Misawa's factories are organized flow line-like with some sections, such as final assembly lines, based on production lines (see also Figure 5.24). In the past, Misawa has distributed its production over nearly 30 factories to be able to address local markets and maintain close customer contact. However, in recent years, production was concentrated on a reduced number of factories. Figures 5.39 to 5.41 outline key steps of Misawa's manufacturing system.

Other Specialties

In the past, Misawa Homes was well known for its pioneering spirit. It was among the first companies in the Japanese prefabrication industry to introduce a quality guarantee system (1966), a school that taught industrialized construction methods (1970), and an after sales and maintenance system (1972). Since 2007, Toyota Home holds a considerable number of shares in Misawa and, in return, has helped to restructure the company. Both companies cooperate in the development of solutions for energy-efficient homes. Furthermore, since 1969 Misawa Homes conducted several experiments in utilizing the possibility of factory eternal logistics by aerial systems (see also **Volume 1**, **Section 4.3.3**) for delivering capsule houses and other material by air transport.

5.4.7 Mitsui Home (Wood Panels)

Product structure	Buildings are composed of wooden panels (Figures 5.42 and 5.43).
Manufacturing layout	Factory is divided into different sections; final assembly of the wooden panels is organized along an assembly line.
Material flow	Flow-line–like material flow.
On-site assembly	Off-site manufacturing takes about 1 day and on-site assembly 2–4 days.

Figure 5.42. Automated fixation and nailing of wood frames. (Photo: Mitsui Home) [This automated assembly line is not currently in operation.]

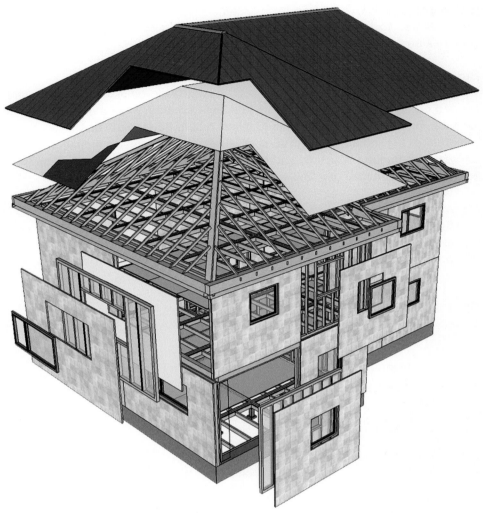

Figure 5.43. Explanation of Mitsui's building kit. (Drawing by W. Pan)

History
The prefabrication company Mitsui was founded in 1974. It originates from, and still belongs to, Mitsui Fudosan, a company that is involved in development, planning, and construction of office buildings, retail buildings, housing, care homes, and hospitals. For more detailed information, see Mitsui (2015).

Business Volume
Mitsui Home has a yearly turnover of about €2.1 billion. It currently sells about 5200 prefabricated houses per year.

Employees and Factories
Mitsui Home operates seven factories located in the Osaka area and has about 2300 employees.

Range of Products

Mitsui Home focusses solely on wood-based buildings and offers two-by-four construction–based buildings as well as wood-based panel kits.

Business Strategy

Mitsui Home targets the upper market segment and offers outstanding design, quality, and performance. Accordingly, its buildings are among the most costly in the Japanese prefabrication industry and highest cost in house makers. Customer services related to planning/customization as well as warranty and maintenance services are of a high quality. Warranty and maintenance services are granted for up to 60 years with the purchase of a Mitsui building.

Manufacturing Strategy

In Mitsui's factory, the production process is generally organized in a flow-line–like manner. Different components, such as panels or roof sections, are in general produced on separate lines. The final assembly is done on the construction site. Figure 5.44 outlines key steps of Mitsui's manufacturing process.

Other Specialties

Mitsui is well known for its outstanding customer services and relations.

5.4.8 Tama Home (Wooden Frame Combined with Panels)

Product structure	The building kit consists of wooden frames and wooden panels (Figure 5.45).
Manufacturing layout and material flow	Subcontracted
Material flow	Subcontracted
On-site assembly	Off-site manufacturing takes about 1 month and on-site assembly 2–4 months.

History

Tama Home Corporation was spun off from the real estate, architecture, and civil engineering business of Chikugo Kosan K.K. and was officially founded in 1998. For more detailed information, see Tama Home (2015).

Business Volume

Tama Home has a yearly turnover of about €1.2 billion (170 billion yen). It currently sells about 9000 prefabricated detached buildings per year, along with other real estate business in Japan as well as other Asian countries. Because of the rising demand for affordable houses with modest quality and performance, the number of houses sold per year is increasing steadily.

Employees and Factories

Tama Home has 2784 employees (as of June 1, 2012). Although Tama Home doesn't have their own prefab factory, one of Tama Home's biggest business partners,

a) Precut wood parts

b) Automated assembly of the panels' wood frames: sequencing, positioning, adjusting, and clamping of wood parts

c) Automated fixation and nailing of wood frames

d) Finishing of wood frames on a conveyor belt

e) Processing of panels on the production line

Figure 5.44. Outline of Mitsui's manufacturing process. (Photos: Mitsui Home) [This automated assembly line is not currently in operation.]

g) Storage for finished wood panels

f) Processing of panels on the production line

h) Preparation of wood panels for transport   i) Assembly of wood panels on the site

Figure 5.44 (*continued*)

a) Tama Home wooden structure (Photo: Tama Home)

Figure 5.45. Outline of Tama Home's key features. (Photos: Tama Home)

Murakami Lumber Co., Ltd. located in Suminoe, Osaka City in Japan, produces precut timber for them.

Range of Products
Tama Home produces only wood-based custom design detached buildings composed of a wooden frame and panels.

Business Strategy
Tama Home offers prefabricated buildings that intentionally provide relatively high-quality buildings at the lowest cost possible with unique performance features that differentiate it from other prefabrication companies. Furthermore, to provide the low cost, Tama Home shortens the customization process and the guidance of the user in the planning phase, and multiple design changes by the customer in that phase are avoided by charging the customer for that service. In contrast to, for example, Sekisui Heim or Pana Home also does not overload its buildings with additional

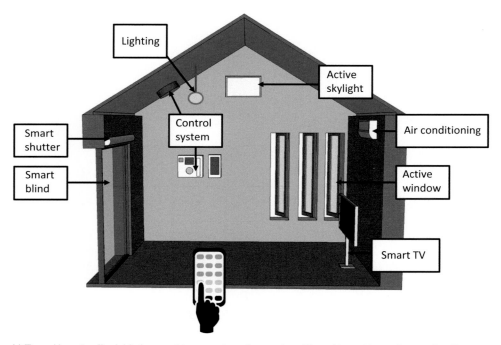

b) Tama Home's affordable low-cost home automation system iTama Home Home Automation System. http://www.tamahome.jp/i-tamahome.

c) Tama Home model house opened in Seijo, Tokyo on April 26th, 2014 to showcase Tama Home's technologies. (Photo: Tama Home)

Figure 5.45 (*continued*)

features (e.g., related to earthquake resistance or smart home technology), and thus manages to provide the customer exactly what he needs at relatively modest price. With this strategy Tama Home recently became a strong competitor of the major prefabrication companies. Furthermore, Tama Home offers a type of lifetime warranty and maintenance service charge free to the customer for 20 years.

Manufacturing Strategy

Tama Home outsources production, to local contractors/builders tha use only domestic materials. Tama Home has not yet built up its own prefabrication factory.

Other Specialties

Tama Home developed a home automation system called iTamaHome and installs that system to its custom-design detached buildings.

5.4.9 Muji House (Wooden Frame Combined with Panels)

Product structure	The building kit consists of wooden frames and wooden panels (Figures 5.46 to 5.48).
Manufacturing layout	Subcontracted
Material flow	Subcontracted
On-site assembly	Off-site manufacturing takes about 1 month and on-site assembly 2–3 months.

History

Muji's housing business is backed by Ryohin Keikaku, who also backs the Muji stores as well as other business operations. For more detailed information, see Muji (2015).

Figure 5.46. Design example. (Photo: Muji House Co., Ltd.)

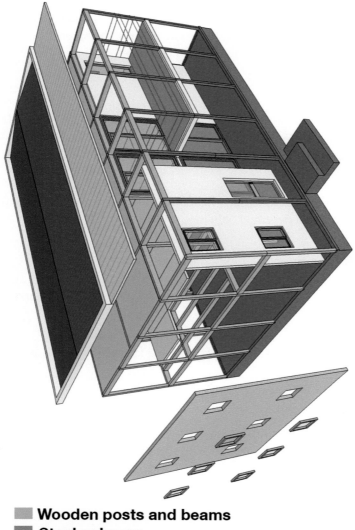

■ Wooden posts and beams
■ Steel columns

Figure 5.47. Explanation of Muji's building kit. (Drawing by W. Pan)

Business Volume
Not disclosed.

Employees and Factories
Muji House, with network partners based on franchise agreements, constructs buildings and sells them. Muji itself has about 50 employees working for its housing business.

Range of Products
Muji House strengthens the framework of their houses by joining pillars and beams with fast-connector systems. Furthermore, similar to other prefabrication companies,

Figure 5.48. View on wood structure and joint. (Photo: Muji House Co., Ltd.)

Muji's housing division intends to boost the sales of products of other divisions that are inbuilt and/or offered along with the buildings. In the case of Muji, its buildings are extensively equipped with compatible Muji furniture. Therefore, room sizes and layouts are fully synchronized with the Muji furniture measurement system (Figure 5.49).

Business Strategy

Muji House targets customers in the mid-30s as purchasers. The heat insulator of the building is composed of external insulation surrounded by rigid urethane foams. Moreover, customer services and customization possibilities are limited in order to keep buildings at a low price and thus affordable for their target customers (limited customization possibilities, only three available basic types [Figure 5.50], warranty and maintenance service limited to 10 years). Similar as, other key players that use prefabricated buildings a platform for selling products of the mother companies backing them (e.g., Sekisui House/Heim: materials and systems made by Sekisui Chemicals, Pana Home: electronic products, smart home technology, solar panels made by Panasonic). Muji House is used by Ryohin Keikaku as platform for selling Muji's furniture and accessories.

Manufacturing Strategy

Muji House subcontracts the production of its kit system as well as the on-site assembly. However, Muji house produces the building infill (furniture, handles, towel rails, mailboxes, etc.) that it embeds into and sells with the building.

Figure 5.49. Synchronization of the building's measurement system with Muji's furniture. (Photo: Muji House Co., Ltd.)

a) Basic Type 1

b) Basic Type 2

c) Basic Type 3

Figure 5.50. Currently available basic types. (Photos: Muji House Co., Ltd.)

Other Specialties
Muji offers buildings with an intentionally simple but good design based on only three standard types.

5.4.10 Sekisui Heim (Steel Units)

Product structure	Buildings are composed of three-dimensional steel-based units (Figure 5.52).
Manufacturing layout and material flow	Consequent organization along an assembly line in main factory. Low depth of added value in the final assembly factory.
Material flow	Uninterrupted material flow, subfactories and suppliers supply parts, profiles, components and appliances JIT or to a small buffer space that flanks the assembly line.
On-site assembly	About 20% of work remains on-site; positioning of units is done by a mobile auto crane within 1–2 days, 4–5 workers necessary on-site during positioning.

History
Sekisui Chemicals Corporation was founded in 1947 and entered the prefabricated housing business around 1970 with the unveiling of its M1 unit-based system. However, Sekisui Chemicals has produced materials and components used in the construction industry since 1947. In 1963, it penetrated the building market further with the introduction of plastic bath tubes. Sekisui Heim, the housing division of

Figure 5.51. Housing units moving on the conveyer belt through phases of the production line. (Photo : Sekisui Heim)

Figure 5.52. Explanation of Sekisui Heim's steel-based building kit. (Drawing by W. Pan)

Sekisui Chemicals, served and still serves as a market platform for a variety of high-performance products (various materials, components and building technologies) of the various divisions of Sekisui Chemicals. For more detailed information, see Sekisui Chemicals (2015).

Business Volume

Sekisui Heim currently has a yearly turnover of about €5 billion and supplies about 15,000 houses per year. In 1997, Sekisui Heim had its highest sales record ever, selling about 35,000 houses.

Employees and Factories

Sekisui Heim operates six factories distributed over Japan and has about 9000 employees. Its average annually employee salary is among the highest in the industry. As Sekisui Heim shifts more work activities from the site into the factories through their unit approach and at the same time automates more, fewer low-skilled workers and more engineers, compared to those employed by competitors, are necessary.

Range of Products

Sekisui Heim focusses on unit-based steel buildings. Since 1984, Sekisui Heim has also supplied unit-based wooden buildings. Usually steel and wooden unit-based buildings were produced in spate factories. However, in the new Hokkaido factory both unit types share the same final assembly line. Furthermore, as a result of the expected growth in demand for wood-based structures, the new factory will allow flexible extension of the factory and enhance its capacity for wooden structures.

| a) Steel profiles before cutting | b) Cutting of steel profiles |

| c) Assisted, manual welding | d) Automatic board plate assembly |

| e) Automated orientation of components | f) Automated welding of components |

Figure 5.53. Key steps of Sekisui Heim's manufacturing process. (Photos: Sekisui Heim)

Business Strategy

Sekisui Heim might provide fewer customized homes than Daiwa House or Sekisui Heim; however, in terms of engineering excellence and quality it can be ranked near the top. This is reflected in the prices for which Sekisui Heim offers its steel unit-based buildings. The average prices for buildings offered are probably the highest in the industry. Sekisui Heim offers relatively compact buildings to a high process, but delivers buildings with outstanding performance and quality. Sekisui Heim

g) Spot welding machine for welding components to components in the process of unit assembly

h) View on the unit assembly and spot welding station

i) Completion of structural unit

j) Electrical wiring

k) Automatic pasting machine

l) Automatic fixation of floor board elements

Figure 5.53 (*continued*)

m) Production of outer wall on a subline   n) Installation of outer wall to unit

o) Installation of inner walls   p) Installation of bath modules

q) Installation of stairs   r) Installation of window frames

Figure 5.53 (*continued*)

s) Installation of partition walls

t) Installation of preconfigured components as sockets and lighting switches

u) Interior finishing line

v) Interior wall finishing

w) Final inspection

x) Completed unit

Figure 5.53 (*continued*)

maintains and extends this position by extensive research, engineering, automated high-precision manufacturing, and rigorous quality control. The company itself claims that the reduced sales in recent years (compared to, e.g., 1997) are part of their strategy that aims at switching into the very top end market segment where higher profit margins can be realized.

z) Installation on the site

y) Shipment from factory to customer's site

Figure 5.53 (*continued*)

Figure 5.54. Part of the production line where the units are equipped with insulation layers. (Photo: Sekisui Heim)

Figure 5.55. Automated installation of components to a wall element on a sub-line. (Photo: Sekisui Heim)

Figure 5.56. Station for robotic welding of frame elements on a sub-line. (Photo: Sekisui Heim)

Figure 5.57. Installation of facade elements to the units on the main production line. (Photo: Sekisui Heim)

Figure 5.58. Robotic quality inspection on the main production line. (Photo: Sekisui Heim)

Manufacturing Strategy

Sekisui Heim, similar to Toyota Home and Misawa (Hybrid) uses three-dimensional steel units that are produced off-site on a production line. Sekisui Heim shifts about 80% of all construction work into the factory, Toyota Home about 85%, and Misawa (Hybrid) about 90%. Sekisui Heim, however, has the most consequently organized manufacturing system (strictly production line–based, see also Figures 5.20–5.23) and fully automated manufacturing process in the factory. About 20% of work remains to be done on-site. This includes positioning of units, which is done by a mobile auto crane within one to two days, and only four or five workers are necessary

Figure 5.59. View of the main production line. (Photo: Sekisui Heim)

on-site during positioning. Figure 5.54 outlines in a sequential manner key steps of Sekisui Heim's manufacturing system. Figures 5.55 to 5.60 outline the utilization of advanced equipment, manipulators, and robots by Sekisui Heim.

Other Specialties

Sekisui Heim was one of the first companies that developed, from about 2007, extremely energy-efficient buildings following the German passive house and plus energy house standards. In 2011, it officially launched the Smart Heim, which is a smart home model series. In addition, Heim cooperates closely with Japanese solar panel manufacturers and has already sold more than 100,000 buildings equipped with solar-panel roofs. Furthermore, Sekisui Heim has successfully established business units in Southeast Asia, and in 2013 opened its new factory in Thailand, which is able to produce around 1000 houses per year.

5.4.11 Toyota Home (Steel Units)

Product structure	Buildings are composed of three-dimensional steel-based units (Figure 5.60).
Manufacturing layout and material flow	Organization in the factory is flow line-like and only partly production line–like. High depth of added value in the final assembly factory (steel profiles are prepared and cut, steel is treated by multiple bathes).
Material flow	Continuous TPS-like material flow. Besides the assembly line, AGVs transporting the units on the factory play an important role.
On-site assembly	About 15% of the work remains on-site; positioning of units is done by a mobile auto crane within eight hours.

History

Toyota Home was founded in 1975 as the housing division of the Toyota Group. In the 1970s, Japan experienced its first robot boom and manufacturing companies applied this technology intensively to the production of nearly every product. Toyota considered its production line–focused Toyota Production System (TPS) as well as its flexible body line (FBL) ideal for producing buildings and had excellent knowledge working with steel (for further information on TPS and FBL, see **Volume 1, Section 4.3**). Toyota Home produces buildings on the basis of three-dimensional steel units. Today, Toyota Home is an independent company with close ties to the Toyota Group. For more detailed information, see Toyota (2015).

Business Volume

Toyota Home currently has a yearly turnover of about €1.5 billion and supplies about 5000 houses per year.

Employees and Factories

Toyota Home operates three factories and has about 3500 employees. Similar to Sekisui Heim, Toyota Home shifts a considerable amount of the construction work activities from the site to the factories, where it creates supportive environments for automation, requiring fewer low-skilled workers and more engineers compared

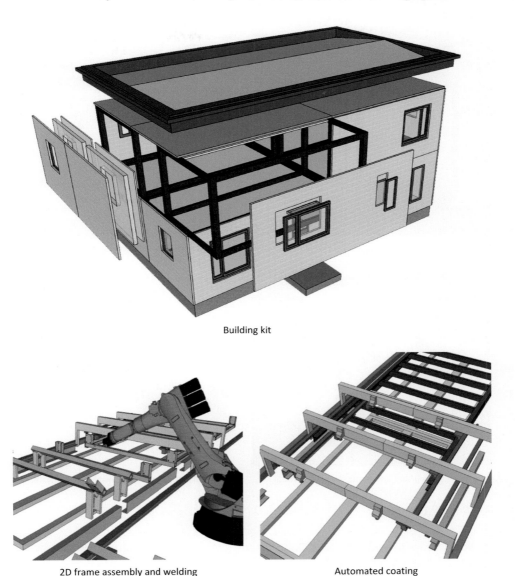

Building kit

2D frame assembly and welding   Automated coating

Figure 5.60. Explanation of Toyota Home's steel-based unit kit. (Drawing by W. Pan)

to competitors. In accordance with the principles of TPS, Toyota Home tailors its assembly process to its workers, who are able to determine the speed of the production line and come up with improvement suggestions.

Range of Products
Toyota Home focusses solely on buildings based on three-dimensional steel units.

Business Strategy
Unlike many of its competitors, Toyota Home does not focus on the very top end sector. However, its buildings are, in terms of quality and cost, still ranked higher than the average conventionally built buildings in Japan. As do other manufacturers,

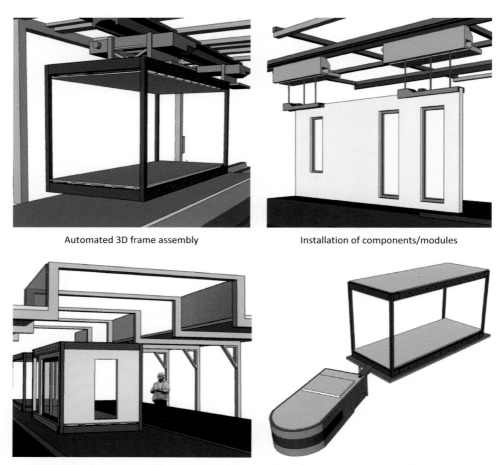

Automated 3D frame assembly

Installation of components/modules

Installation of insulation and finishing on final
assembly line

Use of AGVs as factory internal logistics systems

Figure 5.60 (*continued*)

Toyota Home provides an excellent lifetime warranty and maintenance service for
a time period of up to 60 years.

Manufacturing Strategy

A central element of the manufacturing process is the three-dimensional steel units.
Organization in the factory can be considered as flow line–like. Chain-like and production line–like organization can be found within individual sections. The final
assembly line is strictly production line–based and, in terms of functionality and
design, similar to the final assembly line used in automotive production. The high
depth of added value in the final assembly factory is remarkable an Toyota Home
covers the whole value chain from steel processing to on-site building assembly (in
the factory of Toaota Home: steel profiles are produced and cut, steel is treated and
coated by multiple bathes, steel profiles are assembled/welded into two-dimensional
and further assembled/welded into three-dimensional frames which are then out-
fitted, packed and shipped to the site). A unique characteristic of Toyota Home's
factory is the intensive use of AGVs that are used to move units and components.

Toyota Home has a prefabrication degree of up to 85%, and thus only about 15% of construction work remains on-site; positioning of units is done by a mobile auto crane within eight hours followed by about one month used for minor finishing works.

Other Specialties

In 2004, Toyota Home, together with Toyota Motor Corporation and the Ubiquitous Computing Professor Ken Sakamura, developed the Toyota PAPI house. It is a prototype house embedded with hundreds of microprocessor sensors that can assist the inhabitants. For example, the house can sense vital signs and biorhythms and support the inhabitants in experiencing a deep and relaxing sleep by automatically adjusting the light. The development of the PAPI house can be considered as the transfer of the concept car strategy from automotive industry to construction industry (see also Section 5.1.6). Furthermore, an aim of the experimental building is to explore novel product service system concepts (see also Section 5.5.1).

5.4.12 Misawa Homes Hybrid (Steel Units)

Product structure	Buildings are composed of three-dimensional steel-based units (Figures 5.61 and 5.62).
Manufacturing layout and material flow	Organization in the factory is focussed on a central production main line.
Material flow	The flow of material has a general direction but is, on many points, interrupted in the factory and not continuous.
On-site assembly	About 10–15% of work remains on-site; positioning of units is done by a mobile auto crane within eight hours. Minimum duration of on-site work phase: 30 days, five workers and one crane are necessary.

History

Misawa began operation in 1962 and was initially focused mainly on the prefabrication of wood-based buildings (see also Section 5.4.6). Misawa Homes started its steel home business with the Ceramic Home Hybrid model series in the 1990s. The system is similar to the systems of Sekisui Heim and Toyota Home and is based on the unit approach. For more detailed information, see Misawa (2015).

Business Volume

Misawa Homes supplies up to 1000 unit-based houses per year. Because of recent disasters and the growing demand to build buildings that can withstand strong earthquakes, this number is likely to increase over the next decade.

Employees and Factories

Misawa Homes operates one factory in which it produces its unit-based buildings and employs approximately 500 employees.

Range of Products

Within the Ceramic Home Hybrid model series, customers can choose from a variety of basic exterior designs and floor layouts and fully customize them to fit their requirements. Although the basic element (three-dimensional steel unit) is made of

Figure 5.61. Units for final outfitting on the main assembly line. (Photo: Misawa Homes Hybrid)

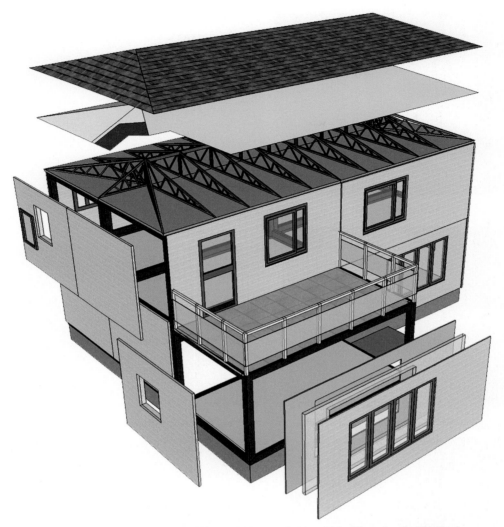

Figure 5.62. Explanation of Misawa Home's steel unit-based kit. (Drawing by W. Pan)

steel, most infill (finishing, partition walls, etc.) is made from wood, Misawa's core business.

Business Strategy

The Ceramic Home Hybrid model series in general does not focus on the very top end sector. The possible top end quality is slightly higher than the quality (and accordingly price range) of Misawa's wood-based buildings. However, its buildings are still higher ranked than the average conventionally built buildings in Japan in terms of quality and cost. Furthermore, Hybrid homes can be delivered in less than 2 months (about 30 days factory production and about 25 days on-site assembly).

Manufacturing Strategy

The production in the factory is focussed on a central production main line in which the three-dimensional steel frames are generated, equipped with finishing,

a) Processing of profiles   b) Manual preassembly

c) Robotic welding   d) Completion of unit frame structure

e) Equipping of unit frame structure   f) Finalization and inspection
with finishing components

Figure 5.63. Misawa's unit-based manufacturing process. (Photos: Misawa Homes Hybrid)

and prepared for transport (see also Figure 5.28). Sublines supply components used to fit these units (exterior walls, interior walls, finishing, appliances, etc.) to the main line. Of all the companies that produce buildings on the basis of steel modules, Misawa has the highest degree of prefabrication and constructs 90% of all necessary construction work in the factory. Figure 5.63 outlines in a sequential manner key steps of Misawa's manufacturing system.

g) On-site assembly step 1

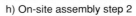

h) On-site assembly step 2

i) On-site assembly step 3

j) On-site assembly step 4

k) On-site assembly step 5

Figure 5.63 (*continued*)

Other Specialties

Owing to its high factory prefabrication degree, Misawa Homes, with its Ceramic Home Hybrid model series, is one of the fastest suppliers of high-quality, customized homes worldwide.

5.4.13 Sekisui Heim Two-U Home (Wood Units)

Product structure	Buildings are composed of three-dimensional steel-based units (Figure 5.64).
Manufacturing layout and material flow	Organization in the factory is flow line–like and only on the final assembly line is it production line–based.
Material flow	Continuous material flow, but less continuous material flow than in steel unit manufacturing.
On-site assembly	The prefabrication ratio is below the 80% average achieved in relation to Sekisui Heim's buildings based on steel units.

History

Sekisui Chemical was founded as a company developing and producing plastics and plastic-based materials in 1947. The business field of Sekisui Chemical grew rapidly owing to the growing use of plastics in almost every business field. Sales of material to the construction business provided an important opportunity, and after obtaining knowledge about the production of construction materials and bath equipment, Sekisui went into the steel house business in the 1970s (see also

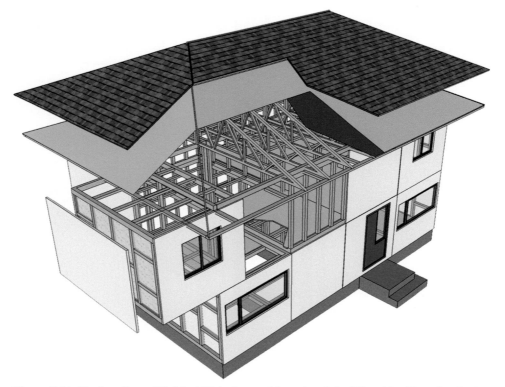

Figure 5.64. Explanation of Sekisui Heim's wood-based unit building kit. (Drawing by W. Pan)

Section 5.4.11). In the mid-1980s Sekisui introduced its Two-U Home wooden pre-fabricated buildings line. Today, Two-U Homes is a sub-brand of Sekisui Heim. For more detailed information, see Sekisui Chemicals (2015).

Business Volume

Sekisui Heim builds on average between 500 and 1000 wood-based unit buildings per year.

Employees and Factories

Sekisui Heim builds wooden units in two of its six factory locations. In one of these locations (Hokkaido), wood and steel units are produced in the same factory and share the same final assembly line.

Range of Products

Aside from buildings based on steel units (which accounts for more than 80% of buildings delivered), Sekisui Heim also delivers buildings based on wooden units (which accounts for less than 20% of buildings delivered). Owing to an increasing ecological awareness the amount of wood-based units, this is expected to grow.

Business Strategy

Sekisui Heim produces wood buildings with nearly the same excellent quality as its steel-based buildings. Because of the slightly lower degree of automation in the wood building manufacturing compared to its steel lines, in general, the same precision cannot be guaranteed. However, in terms of sustainability, wooden buildings offer a multitude of advantages and annual sales are steadily growing.

Manufacturing Strategy

The manufacturing of wooden units is also based on the idea of sending three-dimensional wooden units on an assembly line, where they are equipped with installations, furniture, appliances, insulation, and finishing. However, the tools and processes are different from those of steel-based manufacturing. Furthermore, the organization of the factory layout – in contrast to steel unit manufacturing – is strictly production line–based only at the very end of the manufacturing sequence. The ratio of automation can be considered as lower than in the manufacturing of buildings on basis of steel units (for more information on Sekisui Heims's steel unit manufacturing process, see Section 5.4.11).

Other Specialties

Typically, steel and wood unit manufacturing are housed in clearly divided factory buildings. Sekisui Heim's newest factory in Tohoku is the first factory where wood unit and steel unit manufacturing are unified under one roof and share one common final assembly line (see Section 5.3.3 for a more detailed outline of Sekisui Heim's Hokkaido factory).

5.5 Evolving Tendencies in the Evolution of the Japanese Prefabrication Industry

In the Japanese prefabrication industry, the mere physical product becomes, through the advanced manufacturing process, more and more of a commodity, and companies proceed in the sense of Pine and Gilmore's view on economic evolution (see, e.g.,

Pine & Gilmore, 1999) in the direction of adding and emphasizing complex additional functions and services. The Japanese prefabrication industry will, in the future, deliver much more than the product as an outcome of the construction process. Buildings will become advanced product service systems and the prefabrication companies and their spin-offs will probably deliver a multitude of household and healthcare-related services directly and on-demand to the owner. Furthermore, the Japanese prefabrication industry will, in the future, play a major role in the country's disaster prevention and disaster management strategy; although it does so not out of simple generosity, it will play a major role in Japan's social and cultural system. In close cooperation with major technology manufacturers in Japan (which are actually the parent companies of the prefab industry, see also Section 5.1.5), Japan's prefab industry will soon be able to supply (and in the case of Sekisui House and Daiwa House even to do the real-estate development of the areas on which they are built) perfectly integrated, prefabricated "smart" cities and bundle a multitude of life, mobility and health related services to them. Last but not least, Japan's prefab industry will transfer concepts that it pioneered (e.g., the recustomization approach) towards other industries. These activities will change the notion of buildings as simple construction products and open up completely new value creation possibilities for the currently stagnating Japanese prefabrication industry.

5.5.1 Advanced Product Service Systems

Future concepts in modularity and manufacturing – in particular in general manufacturing industry- aim at bundling/integrating services to/into physical products in order to create novel value creation possibilities and link the customer over time to the manufacturing company/system (see also **Volume 1**, **Sections 4.2** and **4.3**). Service strategies, currently gaining momentum in the Japanese prefabrication industry (see also Linner & Bock, 2012a; on which this section in based on) show that the Japanese prefabrication industry is more acting as a manufacturing industry than a conventional construction industry. The focus in the Japanese prefabrication industry is shifting to "service design" related to the building's life cycle, and similar to, for example, the way smart phones today serve as platforms for services (information, communication, data transfer, and in particular "apps"), prefabricated buildings will integrate an increasing amount of advanced technology for allowing the delivery of a multitude of services throughout the whole product/building life cycle. The traditional distinction between "hard" physical buildings and "soft" household/owner's life-related services will be overcome in favour of the creation of advanced product service systems. This extension of the operational scope gives Japanese prefabrication companies the chance to open up new fields for value creation and to make recurring use of configuration data and built up intense customer relationships (and thus customer confidence/satisfaction as well as knowledge about the customer and recurring purchases of products and services) that last over the whole product/building life cycle. Ultimately this might lead to a scenario where the industry shifts gradually from an economic system of punctual, large payment flows (when a building is purchased) to a system of continuous, small payment flows. Designed in the right way, such a system might have advantages for the prefabrication companies (less risk, steady cash flow, more knowledge about the customer, etc.), their customers (less upfront cost, less cost for loans, increased options, etc.) and the economy as a whole (higher purchasing power).

Physical Services

More and more, Japanese companies are extending their performance focus on "services" related to the building's utilization phase. Here, the companies can use the prefabricated buildings' inherent modularity for upgrades, renovation, rearrangement, and recustomization services. Furthermore, they can offer extraordinary warranty and maintenance services because of the detailed plans and the quality achieved by factory production. Customized energy solutions and personal assistance technologies supporting daily life, health, and handicapped or aged people are emerging building performance fields, which allow prefabrication companies to use and multiply their tremendous knowledge regarding the customer gathered during the configuration or rearrangement procedures.

- *Upgrade services*: The "stock refurbishing business" of Sekisui exemplarily illustrates the service strategy developed by all major Japanese prefabrication companies. Based on detailed plans and data about the delivered housing products, laid down in building information models (BIM), buildings are continuously evaluated. Information about components that should be inspected or changed are generated and reported automatically by a custom software system and forwarded to company staff. The company actively manages the information about delivered products and offers the customer continuously (e.g., every five years) upgrades of interior and exterior design and finishing.
- *Renovation and reorganization services*: Both Sekisui and Toyota offer the possibility to replace or add units owing to changes in lifestyle or household size and demands. Nevertheless, reorganizations can still be simplified and their systems' modularity, standardization, and joining methods offer great potential for continuous rearrangement services. These services could take up and carry on the intense customer relation established through the initial customization process for additional and continuous value creation.
- *Customizable energy platforms*: In 2008, Sekisui Heim started to work with the authors of this chapter on the idea of a customizable, prefabricated, and partly self-sustaining energy and resource platform situated beneath a modular home, a so-called mainboard, inspired by the principle of computing, which accommodates and electronically controls all water instalments and energy components required for a household. The system is designed in a way that allows multiple "mainboard" platforms to interact with each other to form a synergetic relationship between a "mainboard" cluster and its individual components. These platforms can be customized as well as rearranged and can allow for continuous services through the company and cooperating suppliers.
- *Personal assistance technologies*: The Toyota PAPI Dream House, designed and prefabricated based on Toyota Home's cubicle space frames, was co-designed by Ken Sakamura, Professor of Ubiquitous Computing at the University of Tokyo. Courtyard, entrance, kitchen, bathroom, bedroom, and all other rooms have been integrated with sensors, actuators, and assistance technologies to support the inhabitants. The house is a "concept house" showing Toyota Home's

vision of the house of the future. Microsystems have the ability to facilitate household-related services and thus help prefabrication companies to extend their business into the delivery of household-related physical and digital services. Further, Daiwa House cooperates with companies such as Cyberdyne (HAL) and Toto (Intelligent Heath Toilet) to develop assistance technologies and advanced healthcare service systems that can be connected to the prefabricated buildings.

- *Reverse logistics and recustomization:* All buildings of the Japanese prefab company Sekisui Heim, which have been based on a steel frame unit system, can be accepted as trade-ins for a new Sekisui Heim building. Therefore, the deconstruction process and remanufacturing process have been designed as modified reversed versions of the production and assembly process. For deconstruction, first joints between steel frame units are eased, and then the house is transported to a special dismantling factory unit by unit. There, the outdated subcomponents are dismantled and brought into advanced reuse cycles. The bare steel frame units are inspected and subsequently fed again into the production process for customized and user-integrating prefabrication.

Soft and Digital Services

Besides the possibility to customize the building to individual needs (which can also be considered as a kind of service), buildings of Japanese prefabrication companies are distinguished from conventionally built buildings through "soft" and in a next step also "digital" services that accompany the physical product. Currently, "soft" services as, for example, handover services, warranty, and long-term maintenance services comprise standard services accompanying prefabricated buildings in Japan. As most prefabrication companies charge only a very small amount for these services, their main business goal with these services is to attract customers.

Interestingly, all mentioned service models are directly linked to the modular and industrial production of buildings. The celebration of the handover accompanied by a manual illustrates that the house is seen as a product. Warranty and maintenance can be guaranteed for such a long time only because the company has fabricated them under high-quality standards within the factory. Upgrade and renovation services are closely related to the modular design of the buildings by the fact that the company holds the production and customer data generated through the customization and production phase. Finally, almost all major Japanese prefabrication companies are backed by large, financially strong, multinational firms (Sekisui, Toyota, Panasonic, etc.) that are likely to survive the life-cycle period of each of the delivered buildings build a solid and reliable basis for servicing buildings and customers over the (compared to other products as, for example, cars) quite long life cycle of buildings.

The linking of services to products, which Japanese prefabrication companies increasingly focus on, becomes even more interesting when it concerns general developments. More and more business and service concepts aim at creating the basis for product service systems (Sakao and Lindahl, 2009) based on microsystem technology that is then integrated into objects and environments. It is especially necessary to consider the expansion of information and communication technology (ICT) in the field of health and the development towards eHealth – telemonitoring, telecare, teletherapy (Jähn and Nagel, 2004). Primarily, the demographic change, by which Japan will be particularly affected, will necessitate service technologies that are inbuilt in living environments (Wichert and Eberhardt, 2011). In fact, at the moment, many product service systems do not yet aim at custodial or medical services but at services in the field of prevention, fitness, and wellness. Still, this can be seen as an intermediate stage towards integrating telemedical services and services that aim at teleconsultation into the living environment. A large number of the services can be provided digitally, and even (hard) physically performed services (e.g., building rearrangement of recustomization) and the aforementioned soft services (handover, maintenance, etc.) can be supervised digitally to be carried out in a more efficient and high-quality way. Furthermore, services that accompany products provide the possibility to achieve a high degree of individualization/personalization, for example, of a standard product (Chesbrough, 2011). Objects in the living environment that are more and more prevalently equipped with microsystem technology form an ideal interface for initializing and delivering these services (Weber et al., 2005). In general, services that can be digitally initiated/delivered related to the buildings use phase on basis of embedded micro systems technology can be divided into the following categories:

- Classic services in the household for supporting activities of daily life (washing, laundry)
- Security services (theft prevention, fire protection)
- Care service, emergency service
- Services in the field of ageing and care
- Maintenance services
- Services in the field of fitness and health (physiotherapists, doctors, fitness instructors)
- Services for supply of goods and mobility

With the assistance of microsystem technology and data platforms, the efficiency of the provision of these services can be enhanced enormously. Microsystems and data collection platforms can increasingly be established in the living environment (see also **Volume 5** for more information onf future building products embedded with sensors, microsystems, and mechatronics). In terms of the imaginable services this field is not yet fully developed, hence offering the possibility of novel value creation models (see also **Volume 1**, **Chapter 8**). Japanese prefabrication companies have identified this correlation and their activity both in research/development in new service approaches and in broadening their scope of operations (see, e.g., Sanyo Homes, Section 5.4.4, and Pana Home, Section 5.4.3) shows that the shift to a significantly more services-based Japanese prefabrication industry is already on the way.

5.5.2 Prefabrication Industry as Part of a Large-Scale Disaster Management Strategy

As discussed previously (see Sections 5.1.4 and 5.1.12), disasters such as earthquakes, tsunamis, large fires, and typhoons have, throughout the last centuries, again and again created urgent need for large quantities of homes to be supplied within short time periods and created favourable conditions for standardization and the thriving of prefabrication techniques. However, after the Tohoku earthquake, the Japanese prefabrication industry, owing to its capacity and large scale, has actually become a major player in Japan's disaster response strategy. With about 140,000 homes destroyed and about 250,000 people displaced, the supply of shelter and restoration of hygiene, privacy, and other basic functions were key. The Japanese prefabrication industry, in close cooperation with the Japanese government, fulfilled that demand and thus followed contracts and plans made beforehand between the Japanese Prefabrication Construction Suppliers and Manufacturers Association (Prefab Club, 2013) and the government. Today's contribution of the prefabrication industry to the prevention of disasters and the mitigation of the impact caused by disasters is manifold and covers the immediate supply of short-term post-disaster shelters, the rapid production of temporary homes, the integration of advanced damping and earthquake resistance technology in its standard models, and the rapid maintenance and repair service provided for homes damaged by a disaster.

1. *Short-term post-disaster shelters*: The prefabrication industry provides short-term temporary shelters that are in stock and can thus be deployed directly within a few hours following a disaster. Such shelters can serve as homes for individual persons or families and combined as conglomerates for field kitchens, field hospitals, and larger groups of people. With its plans for the EDV-01 (Emergeny Disaster Vehicle Relieve Unit, Daiwalease, 2013; Figures 5.65 and 5.66) the prefabrication industry responded to the lack of appropriate shelter and living conditions in the first month after the Tohoku earthquake. The EDV-01 is a compact, high-tech, and container-like unit that can be transported to the disaster site by any transportation device that can carry containers (e.g., trucks, ships, helicopters). On the site, the unit can automatically transform into a two-floor shelter in which two people can live. The unit contains all basic living functions (water, sewage, gas, communication devices) and even provides electricity and Wi-Fi/Internet connections. The unit is equipped with energy and food supply and functions for about four weeks without the need for connection to any infrastructure.

2. *Temporary housing*: After the Tohoku earthquake and tsunami, the Japanese prefabrication industry, under the lead of the Japanese Prefabrication Construction Suppliers and Manufacturers Association, supplied about 40,000 temporary homes. Temporary homes are defined as homes with an expected life phase of about 10 years. They contain kitchens, toilets, bathrooms, lightning, air conditioning, and gas stoves. They are built on production lines on demand directly after the disaster and are installed to substitute for damaged homes between 1 and 12 months after the disaster.

3. *Advanced dampening and earthquake resistance technology*: In contrast to conventionally built homes, the prefab industry tests prototypes of its models on shaking tables, and thus has brought the general resistance level of its buildings to an outstanding level. Three basic types of measurements can be distinguished:
   Structural resistance (standard equipment): Structural elements and joining systems are optimised for the withstanding of forces caused by earthquakes

Figure 5.65. EDV-1 deployed on the disaster site. (Photo: Daiwa Lease, 2013)

Figure 5.66. EDV-1 taken up by a helicopter for transportation to the disaster site. (Photo: Daiwa Lease, 2013)

and/or typhoons. In terms of earthquake resistance, steel construction, and in particular, the unit construction method used by Sekisui Heim and Toyota Home, has proven to be advantageous. Unit construction provides additional redundancy: even if structural elements from one unit fail, the other surrounding units can still keep the building stable. Structural resistance belongs to the standard requirements of buildings and guarantees that the building does not collapse, no major damages occur to the building, and no elements fall off and cause major injury.

Vibration control panels (optional equipment, Figures 5.67 and 5.68): These are based on rubber dampening technology and viscoelastic elements absorbing the energy in case of an earthquake. Bracings of panels from Sekisui House and Daiwa House are equipped with dampening elements. On each floor, about three such panels should be installed. The panels minimize the impact of the vibrations caused by an earthquake on a building and its interior. In terms of cost, they add between €10,000 and 50,000 (depending on the size of the building and the number of panels to be installed) to the cost of the home.

Base insulation systems (optional equipment; Figure 5.69): These decouple the building from the ground and rest the whole building on a roller bearing system (sliding isolation system). Such systems are expensive and add about €30,000 to 100,000 (depending on the size of the building) to the cost of a home. However, base isolation is the most effective way to avoid the impact of earthquakes on the building, its interior (computers, furniture, etc.), and people (preventing injuries, etc.).

4. *Rapid maintenance and repair service*: Centuries ago, after major disasters, Japanese master builders repaired and maintained damaged buildings. Today this role has been partly taken over by the Japanese prefabrication industry. As mentioned previously, a major reason for the high cost of Japanese

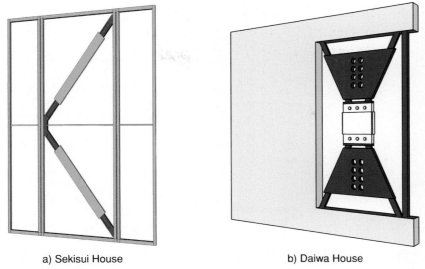

| a) Sekisui House | b) Daiwa House |

Figure 5.67. Earthquake vibration control panels.

prefabricated buildings is that the companies provide the delivery of the home with a kind of lifetime maintenance and repair service. Therefore, after major disasters, the Japanese prefabrication industry reduces its regular operation to be able to fulfil its duty to restore damaged homes and also to inspect homes that are not obviously damaged to guarantee their functioning.

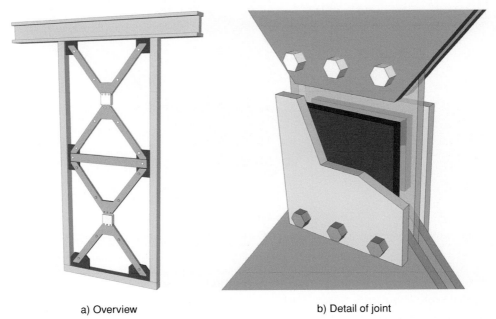

| a) Overview | b) Detail of joint |

Figure 5.68. Earthquake vibration control panel SanDouble-X, Sanyo Homes. (Drawing by B. Georgescu)

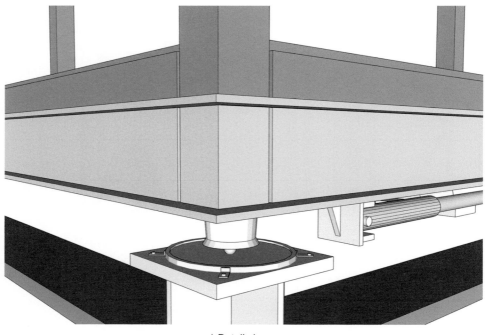

a) Detail view

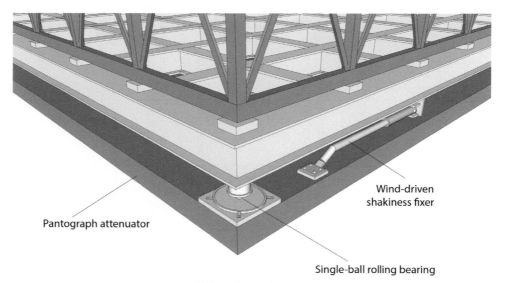

Wind-driven
shakiness fixer

Pantograph attenuator

Single-ball rolling bearing

b) Overview system

Figure 5.69. Seismic base isolation system, Daiwa House. (Drawing by B. Georgescu)

5.5.3 Extending the Value Chain through the Development of Prefabricated, Sustainable High-tech Settlements

Panasonic recently announced plans for its Fujisawa Sustainable Smart Town (Fujisawa SST; Figure 5.70), a settlement with about 1000 housing units that was realized quickly in 2014 and celebrated its opening in November 2014 (see also Fujisawa

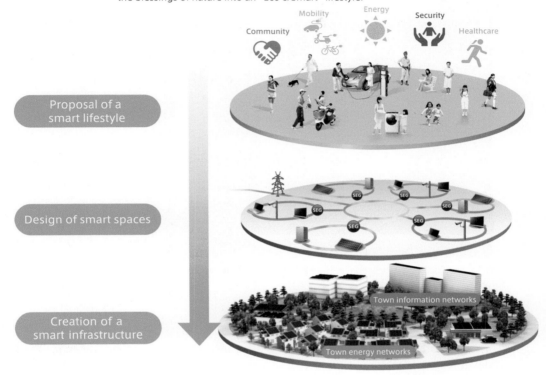

a) The town design of Fujisawa SST is based on residents' lifestyle. It consists of three layers, without excessive emphasis on zoning or infrastructure design. FSST's goal is to create a sustainable smart town that incorporates the blessings of nature into an "Eco & Smart" lifestyle.

Figure 5.70. Fujisawa Sustainable Smart Town. Panasonics Housing Division Pana Home supplied the buildings, and other Panasonic divisions the building technologies, "green" electronics, home appliances, and assistance technologies. (Photos: Pana Home)

SST, 2015). Panasonic's Housing Division, Pana Home, will supplied the buildings, and other Panasonic divisions the building technologies, "green" electronics, home appliances, and assistance technologies, as, for example, Panasonic's robotic bed for the elderly and disabled. Other smaller, but still quite large settlements are planned based on the same approach (e.g., Pana Home's Smart City Shioashia). Through this approach, Panasonic has put itself in a developer position (similar to that of a super-OEM) utilizing its own subdivisions for the supply of nearly everything that comprises a city (including building/city life-cycle services related to security, mobility health care and energy). Panasonic (with the help of Pana Home) therefore is extending its reach over almost the entire value chain. Similarly, Daiwa House, Sekisui Chemical (with its related companies Sekisui House and Sekisui Heim) and

b) Panasonic's community solar panels will be integrated in surfaces and roofs.

Figure 5.70 (*continued*)

also Sanyo Homes have done the same thing increasingly over the last decade. The developer approach can be considered a good measurement against stagnation and, during the last decade, even slightly decreasing demand for prefabricated buildings. It boosts both the sales of the housing division and other divisions.

The developer approach also gives Panasonic the ability to innovate and use the settlement as a test bed for new technologies and social concepts. The Fujisawa SST is an actual operational town. The construction consortium included 17 firms and 1 association; Panasonic: Pana Home Corporation; Dentsu Inc.; Tokyo Gas Co., Ltd.; Nippon Telegraph and Telephone East Corporation; Mitsui Fudosan Co., Ltd.; Mitsui Fudosan Residential Co., Ltd.; Mitsui & Co., Ltd.; Culture Convenience Club Co., Ltd.; So-Two Inc., Sunautas Co., Ltd.; Ain Pharmaciez Inc.; Nihon Sekkei, Inc.; Accenture; Gakken Cocofump Holdings Co., Ltd.; Sohgo Security Services Co., Ltd.; & the social welfare corporation Nagaoka.

Together with Tokyo Gas, Panasonic also developed its home fuel cell Ene-Farm (Figure 5.71), which has been produced (Panasonic) and sold (Tokyo Gas) since 2011 in an upgraded version. The fuel cell is a "co-generation" system producing energy (electricity and heat) as an outcome of a chemical reaction of oxygen (retrieved from the atmosphere) and hydrogen produced from the gas supplied by Tokyo Gas or other gas suppliers. The fuel cell is super-compact and is only the size of a wardrobe. Pana Home buildings can (optionally) integrate the fuel cell and are designed to optimize its performance.

Figure 5.71. Panasonic's super-compact Ene-Farm home fuel cell. (Photo: Pana Home)

Together with the University of Tokyo's Institute for Robot Technology (IRT), Panasonic also researches robotic solutions to be incorporated into the kitchens of its Pana Home buildings. One of the first projects was a robot that supports activities such as dishwashing and cooking.

Another concept aimed at sustainability, in terms of closed loop and short-way resource supply, is taken further by Daiwa House's Agri-Cube (Figure 5.72), a compact and prefabricated unit that incorporates the latest technology for hydroponic vegetable "production". The unit can be installed in detached houses, gardens, neighbourhoods, office buildings, rooftops of restaurants, and within high-rise buildings. Furthermore, Agri-Cubes can also be used as temporary food supply modules following disasters until the local infrastructure is restored. Daiwa House has already built prototypes of the unit and is currently working on the marketing strategy for it. The unit will probably cost about €50,000 and the cost for its operation is estimated at €3000 to 4000 per year.

Similarly to Panasonic, Daiwa House is opening up new market fields and adding value to its products by incorporating emerging technologies. It does so not only with the Agri-Cube, but also in the field of emerging building integrated technologies. Together with the bath module supplier Toto, Daiwa House developed an intelligent toilet that is able to measure vital signs and thus monitor the health status of each family member. The toilet allows for each family member to create his or her own account, track his or her records and send – if wanted – the data directly to a doctor. In particular, in the face of changing demographics and ageing societies, technological advances in such assistance systems are currently a major topic in Japan. Daiwa House has also cooperated with Prof. Sankai and Cybedyne Corporation in the

a) Outside view of the unit which is designed within the container measurement system and can be transported by a standard truck.

b) The unit incorporates technology for hydroponic vegetable "production".

Figure 5.72. Agri-Cube, a compact and prefabricated indoor farming unit. (Photos : Daiwa House)

c) The unit is easy to transport and install in place via crane.

Figure 5.72 (*continued*)

integration of the robotic exoskeleton assistive suite (Hybrid Assistive Limb) as a subsystem into its buildings.

With its developments in the field of farming units and assistance systems, Daiwa House has strengthened both its image of working on the forefront of technological advances as well as its future market and value creation flexibility.

5.5.4 Reverse Innovation: Mass-Customized Housing Production as a Prototype for Future Manufacturing Systems

It is often discussed that advanced manufacturing technologies applied in other industries, such as the automotive industry, should be transferred to or used in building construction. The Japanese prefabrication industry, however, has reached, in terms of strategies, methods, and technologies, a level that actually allows knowledge transfer in the other direction: from the construction industry to other industries that deal with the production of complex products (for more information on the idea of reverse innovation, see also **Volume 1, Section 7.5.2**). A manufacturing technology as large scale and automated as prefabrication in Japan that is able to mass-customize/personalize highly complex products such as buildings in a near real-time and OPF manner (see Section 5.3) can be used for individual, automatic, and OPF-like manufacturing, as is the case with the manufacturing of other complex products such as cars, aircraft, or ships. Overcoming the challenge of automatically manufacturing buildings could lead to concepts and technologies that are of great interest also for the manufacturing industry as a whole.

As discussed in **Volume 1, Section 7.5.3**, at about €80,000 to 100,000 the Phaeton is relatively inexpensive for a "handmade" luxury car. However, it is still a very high-end, luxury car, and the Phaeton cannot be customized at the same price level as a standard VW car. Indeed, it costs 200 to 300% more than a standard VW. On the contrary, Japanese LSP companies allow more personalized products (that allow for intense and flexible degrees of customer integration in design, product engineer and life-cycle related servicing; see, e.g., Section 5.2.9) to be produced at nearly the same, or at only a slightly higher cost as conventional buildings (depending on the customer's wishes), up to a maximum of 50% more. The basis for this ability is the automation of low-level component production in combination with a perfected and highly flexible OPF organization, alongside a single production line. ERP systems for controlling the production and logistic flow were introduced and refined for this purpose in the 1970s. All in all, it can be said that the Japanese LSP industry meets the requirements of real and affordable mass customization (Piller, 2006), more so or better than the automotive industry. Both Sekisui Heim and Toyota Home have advanced the concept of the TPS that they adopted in the 1980s (see Sections 5.1.9 to 5.1.12) to a demand-oriented manufacturing system that by far exceeds the current ability to mass customize in automotive manufacturing.

Because eco-efficiency is a thriving topic in the general manufacturing industry, the concept of zero-waste factories and recustomization already realized by Japanese companies (see Section 5.3.2) could be inspiring for the automotive, aircraft, or shipbuilding industries. Sekisui Heim and Toyota Home show that industrialized and highly automated factory production of buildings can address multiple sustainability aspects. Resource consumption is reduced through demand-oriented manufacturing, structured factory environments, closed loop recycling, and energy harvesting within manufacturing systems. Both have already set up zero-waste factories. Furthermore, by a recustomization approach, all obsolete building modules of Sekisui Heim can be accepted as trade-ins towards a new Sekisui Heim building. Therefore, the deconstruction process is a reversed and modified version of the construction process, as it is based on subsequent unit factory completion of modular units on the conveyor belt.

More information about reverse innovation and the transfer of advanced concepts and technologies from advanced off-site and on-site building manufacturing approaches to other industries is given in **Volume 1, Section 7.5.3.**

References

Abdi, O., Kowalsky, M., Hassan, T., Kiesel, S., Peters, K. (2008) "Large deformation polymer optical fiber sensors for civil infrastructure systems". In *Proceedings of the Conference on Sensors and Smart Structures Technologies for Civil, Mechanical, and Aerospace Systems*. San Diego: The International Society for Optical Engineering.

Abley, I., Schwinge, J. (2006) *Manmade Modular Megastructures*. Hoboken, NJ: John Wiley-A & Sons.

Aicher, H., Reinhardt, W., Garrecht, H. (eds.) (2013) *Materials and Joints in Timber Structures: Recent Developments of Technology*. New York and Heidelberg: Springer Science+Business Media.

Allen, E., Iano, J. (2009) *Fundamentals of Building Construction: Materials and Methods*. Hoboken, NJ: John Wiley & Sons.

Altintas, Y. (2012) *Manufacturing Automation: Metal Cutting Mechanics, Machine Tool Vibrations, and CNC Design*. Cambridge: Cambridge University Press.

Andres, J., Bock, T., Gebhardt, F., Steck, W. (1994) "First results of the development of the masonry robot system ROCCO: A fault tolerant assembly tool". In *Proceedings of 11th International Symposium on Automation and Robotics in Construction*, pp. 87–93. Brighton, UK: International Association for Automation and Robotics in Construction.

Asahi Kasei (2015) http://www.asahi-kasei.co.jp/asahi/en/aboutasahi/history/ (Accessed 12 February 2015).

Atterbury, G. (1920) U.S. Patent No. 1326902. Washington, DC: U.S. Patent and Trademark Office.

Autopullit. http://www.autopulit.com/en/ (Accessed 17 October 2014).

Baufritz. Wir bauen Werte seit 1896. www.baufritz.d (Accessed 19 November 2013).

Bauser, M., Sauer, G., Siegert, G. (2006) *Extrusion*. Materials Park, OH: ASM International.

Bechthold, M., King, J., Kane, A., Niemasz, J., Reinhart, C. (2011) "Generation of a prototypical high performance ceramic shading system". In *Proceedings of 28rth International Symposium on Automation and Robotics in Construction*, pp. 70–75. Seoul, South Korea: International Association for Automation and Robotics in Construction.

Bergdoll, B., Christensen, P., Broadhurst, R. (eds.) (2008) "Home delivery: Fabricating the modern dwelling". New York: The Museum of Modern Art.

Bien-Zencker. Über Bien-Zenker, Häuser, Bauweise. www.bien-zenker.de (Accessed 19 November 2013).

Birch, A. (2005) *Economic History of the British Iron and Steel Industry*. Oxon, UK: Routledge.

Bock, T. (2008) "Digital design and robotic production of 3D shaped precast components". In *Proceedings of 25th International Symposium on Automation and Robotics in Construction*, pp. 11–21. Vilnius, Lituania: International Association for Automation and Robotics in Construction.

Bock, T., Linner, T. (2009a) "Automation and robotics in on-site production and urban mining". In G. Girmscheid, F. Scheublin (eds.), *New Perspective in Industrialization in Construction – A State-of-the-Art Report*, pp. 281–298. Zurich: CIB Publications.

Bock, T., Linner, T. (2009b) "Customized automation and robotics in prefabrication of concrete elements". In G. Girmscheid, F. Scheublin (eds.), *New Perspective in Industrialization in Construction – A State-of-the-Art Report*, pp. 207–231. Zurich: CIB Publications. (This publication in extensively rewritten and expanded form built the basis for Section 2.2.)

Bock, T., Linner, T. (2009c) "Customized automation and robotics in prefabrication of wood eElements". In G. Girmscheid, F. Scheublin (eds.), *New Perspective in Industrialization in Construction – A State-of-the-Art Report*, pp. 263–280. Zurich: CIB Publications. (This publication in extensively rewritten and expanded form built the basis for Section 2.3.)

Bock, T., Linner, T. (2009d) "From early trials to advanced computer integrated prefabrication of brickwork". In G. Girmscheid, F. Scheublin (eds.), *New Perspective in Industrialization in Construction – A State-of-the-Art Report*, pp. 161–181. Zurich: CIB Publications. (This publication in extensively rewritten and expanded form built the basis for Section 2.1.)

Bock, T. (2011a) "Karakuri Kultur in Architektur und Bauwesen". In *Mensch-Roboter-Interaktionen aus interkultureller Perspektive- Japan und Deutschland im Vergleich*, pp. 18–32. Veröffentlichungen des Japanisch-Deutschen-Zentrums Berlin, Band 62. Berlin: Japanisch-Deutsches-Zentrum.

Bock, T., Linner, T., Miura, S. (2011b) "Robotic high-rise construction of pagoda concept: innovative earthquake-proof design for the Tokyo sky tree". In *Proceedings of Council for Tall Buildings and Urban Habitat (CTBUH) World Conference*, pp. 659–669. Seoul, South Korea: CTBUH.

Bonwetsch, T., Kobel, D., Gramazio, F., Kohler, M. (2006) "The informed wall: Applying additive digital fabrication techniques on architecture". In *Proceedings of the 25th Annual Conference of the Association for Computer-Aided Design in Architecture*, pp. 489–495. Louisville, KY.

Bosche, F., Haas, C. T. (2008) "Automated 3d data collection (A3DDC) for 3D building information modelling". In *Proceedings of the 25th International Symposium on Automation and Robotics in Construction*, pp. 279–285. Vilnius, Lituania, International Association for Automation and Robotics in Construction.

Bowley, B. (1994) "Calcium silicate bricks". *Structural Survey*, 12(6):16–18.

Broad (2015) http://www.broad.com/ (Accessed 12 February 2015).

Brzev, S., Guevara-Perez, T. "Precast concrete construction". World Housing Encyclopedia Reports. www.world-housing.net/wp-content/uploads/2011/08/Type_Precast.pdf: 32 (Accessed 19 November 2013).

Campbell, J. (2011) *Complete Casting Handbook: Metal Casting Processes, Metallurgy, Techniques and Design*. Oxford: Elsevier.

Carbajal, L., Rubio-Marcos, F., Bengochea, M. A., Fernandez, J. F. (2007) "Properties related phase evolution in porcelain ceramics". *Journal of the European Ceramic Society*, 27(13):4065–4069.

Caristan, C. L. (2004) *Laser Cutting Guide for Manufacturing*. Dearborn, MI: Society of Manufacturing Engineers.

Carter, C. B., Norton, M. G. (2013) *Ceramic Materials*. New York: Springer Science+Business Media.

CEMBUREAU. www.cembureau.eu (Accessed 22 November 2013).

Chawla, K. K. (1998) Processing of ceramic matrix composites. In *Ceramic Matrix Composites*, pp. 212–251. New York: Springer-Verlag.

Chesbrough, H. (2011) *Open Services Innovation*. San Francisco: Jossey-Bass, John Wiley & Sons.

Chugg, W. A. (1964) *Glulam: The Theory and Practice of the Manufacture of Glued Laminated Timber Structures*. London: Benn.

Cicek, T., Tanriverdi, M. (2007) "Lime based steam autoclaved fly ash bricks". *Construction and Building Materials*, 21(6):1295–1300.

Concept-Laser. http://www.concept-laser.de/ (Accessed 17 October 2014).

Council for Construction Robot Research (1999) "Construction robot system catalogue in Japan". Research report. Tokyo: Robot Association.

Csanády, E., Magoss, E. (2012) *Mechanics of Wood Machining*. New York and Heidelberg: Springer Science+Business Media.

Daiwa House (2015) http://www.daiwahouse.co.jp/English/history/ (Accessed 12 February 2015).

Daiwalease. Daiwa House Leasing and Renting Division. http://www.daiwalease.co.jp/edv-01/edv01_concept.html (Accessed 19 November 2013).

Danobat Machinery. http://www.danobatgroup.com/en/danobat (Accessed 17 October 2014).

Department for Communities and Local Government (DCLG) (2013) "English housing survey 2011 to 2012 – headline report". London: National Statistics, United Kingdom.

Dorean, D. K., Carthe, R. (1992) *Construction Materials Reference Book*. Oxford: Butterworth Heinemann.

Dunne, T., Dietrich, W. E., Humphreyand, N. F., Tubbs, D. W. (1980) "Geologic and geomorphic implications for gravel supply". In *Proceedings of the Conference Salmon-Spawning Gravel: A Renewable Resource in the Pacific Northwest?, Seattle*.

Dupont Lighstone. http://www.dupontlightstone.com (Accessed 17 October 2014).

Eastman, C., Teicholz, P., Sacks, R., Liston, K. (2011) *BIM Handbook: A Guide to Building Information Modeling for Owners, Managers, Designers, Engineers and Contractors*. Hoboken, NJ: John Wiley & Sons.

Eco-serve. http://www.2020-horizon.com/ECO-SERVE-European-construction-in-service-of-society(ECO-SERVE)-s36879.html (Accessed 17 October 2014).

Elkmann, N., Felsch, T., Forster, T. (2010) "Robot for rotor blade inspection". In 1st International Conference on, Applied Robotics for the Power Industry (CARPI), pp. 1–5, 5–7 October. doi: 10.1109/CARPI.2010.5624444.

EN 10020. Directive number of the Eurpean Union: 07 of the year 2000.

ERMCO. http://www.ermco.eu/documents/ermco-documents/statistics.xml?lang=en (Accessed 15 June 2014).

FAO (2009) Food and Agriculture Organization of the United Nations, "State of the World's Forests 2009 – Global Demand for Wood Products". Report.

Fujimoto, T. (1999) *The Evolution of a Manufacturing System at Toyota*. Oxford and New York: Oxford University Press.

Fujisawa SST (2015) http://fujisawasst.com/EN/ (Accessed 12 February 2015).

Furuse, J., Katano, M. (2006) "Structuring of Sekisui Heim automated parts pickup system (HAPPS) to process individual floor plans". In *Proceedings of 23rd International Symposium on Automation and Robotics in Construction*, pp. 352–356. Tokyo: International Association for Automation and Robotics in Construction.

Gagnon, S., Pirvu, C. (eds.) (2011) *CLT Handbook: Cross-Laminated Timber*. Pointe-Claire, Canada: FPInnovations.

Geinsa. http://geinsa.com/en/ (Accessed 17 October 2014).

Girmscheid, G. (2010) "Off-site production methods – precast plant production". In G. Girmscheid, F. Scheublin (eds.), *New Perspective in Industrialisation in Construction, A State of the Art Report*, pp. 233–261. Zurich: ETH/CIB Publications.

Góngora, R. (2013) "Challenges in early ship design: integrated solution from concept to production". In *Proceedings of International Conference on Computer Applications in Shipbuilding ICCAS-2103*, Busan, Korea.

Gunßer, C. (2002) "Schnelles Haus, schönes Haus". In K. Tschavgova (ed.), *Zuschnitt 6 Vor fertig los*, pp. 5–7. Vienna: Proholz Austria.

Händel, F. (2007) *Extrusion in Ceramics*. New York and Heidelberg: Springer Science+Business Media.

Hanser, A. (2002) "Vorfertigung im internationalen Vergleich". In K. Tschavgova (ed.), *Zuschnitt 6 Vor fertig los*, pp. 8–10. Vienna: Proholz Austria.

Harvey, P. D. (1982) *Engineering Properties of Steel*. Geauga County, OH: American Society for Metals.

Hirose, M., Y. Hajime, K. Seng, T. Saikaku, N. Yasufumi, I. Souhei, K. Kisho, N. Yoshikazu, M. Kiwa, Y. Yoshiyuki, H. Naohiko (2011) *Metabolism, the City of the Future* [Exhibition catalogue]. Tokyo: Mori Art Museum.

Ho, P. (2005). *Institutions in Transition: Land Ownership, Property Rights, and Social Conflict in China.* Oxford Scholarship Online.

Hollow, M. (2011). "Suburban ideals on England's interwar council estates". *Journal of the Garden History Society*, 39(2):203–217.

HufHaus. Das Unternehmen, Standorte, Portfolio. www.huf-haus.com (Accessed 19 November 2013).

Jähn, K., Nagel, E. (2004) '*eHealth*'. Berlin: Springer Science+Business Media.

Jeffus, L. (2011) *Welding: Principles and Applications.* Clifton Park, NY: Delmar Cengage Learning.

Kahn, M. (ed.) (1986) *Fine Woodworking on Wood and How to Dry It.* Newtown, CT: Tauton Press.

KAMPA. Unternehmen, Häuser Vielfalt, Innovative Bauweise. www.kampa.de (Accessed 19 November 2013).

Kawano, A., Sakino, K., Kuma, K., Nakata, K. (2002) "Seismic resistant system of multi-story frames using concrete-filled tubular trusses". In *Proceedings of the Twelfth International Offshore and Polar Engineering Conference, Kitakyushu (Japan),* The International Society of Offshore and Polar Engineers.

Kawazoe, Y. (1960) *Metabolism/1960.* Tokyo: Toppan Printing Co. (Reprinted and re-edited 2011 by Echelle).

Khamidi, M. F., Matsufuji, Y., Yamaguchi, K. (2004) "Adapting dry-masonry brick house system as a green cycle model for South East Asian markets". In *International RILEM Conference on the Use of Recycled Materials in Buildings and Structures*, Barcelona, Spain.

Khan, A., Lemmen, C. (2013) "Bricks and urbanism in the Indus Valley rise and decline". arXiv preprint arXiv:1303.1426.

Knight, W. A., Boothroyd, G. (2005) *Metal Machining and Machine Tools.* Boca Raton, FL: CRC Press.

Kolb, J. (2009) Der kleine Schritt von Vorfertigung zum Fertighaus. *ARCH+*, 193:36–37.

Krenkel, W, Berndt, F. (2005) "C/C–SiC composites for space applications and advanced friction systems". *Materials Science and Engineering*, 412(1–2):177–181.

Kües, U. (ed.) (2007) *Wood Production, Wood Technology and Biotechnological Impacts.* Göttingen, Germany: Universitätverslag Göttingen.

Kuhnert, N. (2010) Smart Price Houses. *ARCH+*; 198/199:20–25.

Lantek. http://www.lanteksms.com/uk/ (Accessed 17 October 2014).

Lee, J., Chang, B. C., Lee, S., Bernold, L. B., Lee, T. S. (2011) "Automa system for lunar landing pad". In *Proceedings of 28th International Symposium on Automation and Robotics in Construction*, pp. 975–980. Seoul, South Korea: International Association for Automation and Robotics in Construction.

Lenke, P. Wendt, M., Krebber, K., Seeger, M., Thiele, E., Metschies, H., Gebreselassie, B., Munich, J. C. (2009) "Polymer optical fiber sensors for distributed strain measurement and application in structural health monitoring". *IEEE Sensors Journal*, 9:1330–1338.

Lim, S., Buswell, R. A., Le, T. T., Austin, S. A., Gibb, A. G. F., Thorpe, T. (2012) "Developments in construction-scale additive manufacturing processes". *Automation in Construction.* 21:262–268.

Lim, S., Buswell, R., Le, T., Wackrow, R., Austin, S., Gibb, A., Thorp, T. (2011) "Development of a viable concrete printing process". In *Proceedings of 28thInternational Symposium on Automation and Robotics in Construction*, pp. 665–670. Seoul, South Korea: International Association for Automation and Robotics in Construction.

Linner, T. (2013) *Automated and Robotic Construction: Integrated Automated Construction Sites.* Dr.-Ing. dissertation, Technische Universität München.

Linner, T., Bock, T. (2009) "Smart customization in architecture: towards customizable intelligent buildings". In *Proceedings of Conference on Mass Customization, Personalization and Co-creation (MCPC 2009)*, Helsinki, Finland: MCPC.

Linner, T., Bock, T. (2012a) "Evolution of large-scale industrialization and service: Innovation in Japanese prefabrication industry". *Journal of Construction Innovation: Information, Process, Management*, 12(2):156–178. Sections 5.1, 5.2, and 5.3 of this volume are partly based on/inspired by this journal article. The included ideas and text sections have been extensively rewritten and expanded for this volume.

Linner, T., Bock, T. (2012b) "Demographic change robotics: Mechatronic assisted living and integrated robot technology". In N. McDowall, D. Chugo, S. Yokota (eds.), *Introduction to Modern Robotics II*, pp. 19–46. Hong Kong: iConcept Press.

Linner, T., Bock, T. (2013) "Automation, robotics, services evolution of large-scale mass customization in the Japanese building industry". In A. E. Piroozfar, F. T. Piller (eds.), *Mass Customisation and Personalisation in Architecture and Construction: A Compendium of Customer-centric Strategies for the Built Environment*, pp. 154–163. London & New York: Routledge/Taylor & Francis Group.

Loire Safe Automated Presse. http://www.loiresafe.com/index.php?lang=en/ (Accessed 17 October 2014).

Loison-Leruste, M., Quilgars, D. (2009) "Increasing access to housing: Implementing the right to housing in England and France". *European Journal of Homelessness*, 3:75–100.

Luccon. http://www.luccon.com (Accessed 17 October 2014).

Luo, X. (2004). "The role of infrastructure investment location in China's Western development". World Bank Policy Research Working Paper 3345. Washington, DC: World Bank.

Lye, P. F. (1985) *Woodwork Theory*, Book 3. Cheltenham, UK: Nelson Thornes.

Maas, P., Peissker, P., Ahner, C. (2011) *Handbook of Hot-dip Galvanization*. Hoboken, NJ: John Wiley & Sons.

Maekawa, K., Obikawa, T., Yamane, Y., Childs, T. H. C. (2000) *Metal Machining: Theory and Applications*. Oxford: Elsevier.

Matsukuma, H., Inada, T., Yuzuhana, A., Tani, T., Matsuzawa, H., Iachida, T. (2006) *Kuino Mayekawa Retrospective*. [Exhibition catalogue]. Tokyo: The Japan Institute of Architecture.

Matsumura, S. (1997) "The composition of the house-building market in Japan and its trends". In *Proceedings of the Scandinavia-Japan Seminar on Future Design and Construction in Housing*, pp. 13–24. Stockholm, Sweden.

Matsumura, S. (2010) "Japanese prefabrication industry" Presentation/lecture given at The University of Tokyo.

May, M. (2006) *IT im Facility Management erfolgreich einsetzen*. Berlin: Springer Science+Business Media.

Meier, R., Piller, F. T. (2011) "Systematisierung von Strategien zur Individualisierung von Dienstleistungen". Research report. Arbeitsberichte des Lehrstuhls für Allgemeine und Industrielle Betriebswirtschaftslehre an der Technischen Universität München.

Milgrom, P., Roberts, J. (1990) "The economics of modern manufacturing: technology, strategy and organization". *American Economic Review*, 80(3):511–528.

Misawa (2015) http://www.misawa.co.jp/en/info/history.html (Accessed 13 February 2015).

Mitsui (2015) http://www.mitsuifudosan.co.jp/english/corporate/about_us/history (Accessed 12 February 2015).

Morimoto, N., Kobayashi, H., Matsumura, S. (1994) "Houses and technologies". *The Book for Understanding Japanese Houses*, pp. 113–151. Tokyo, Japan: Japan Housing Loan Corporation, PHP Research Institute.

Moses, D. M., Prion, H. G. L., Li, H., Boehner, W. (2003) "Composite behavior of laminated strand lumber". *Wood Science and Technology*, 37(1):59–77.

MTorres. http://www.mtorres.es// (Accessed 17 October 2014).

Muji (2015) http://www.muji.net/ie/ (Accessed 12 February 2015).

Murphy, D. (2009) *Moshe Safdie*, Vol. 1. Victoria, Australia: Images Publishing.

Nagata, F., Watanabe, K., Kiguchi, K. (2006) "Joystick teaching system for industrial robots using fuzzy compliance control". In S. Cubero (ed.), *Industrial Robotics:Theory, Modelling and Control*, pp. 362–367. Rijeka, Croatia: Intech, ProLiteratur Verlag, Germany/ARS, Austria.

Nawy, E. G. (2008) *Concrete Construction Engineering Handbook*. Boca Raton, FL: Taylor & Francis Group.

Neison, A. (2010) *Practical Wooden Boat Building*. Bremen, Germany: Salzwasser-Verlag.

Nikkei BP. "Securities report for investors in financial tear 2009". Research report. http://www.nikkeibp.co.jp (Accessed 10 September 2012).

Noguchi, M. (2013) "Commercialisation principles for low-carbon mass-customized housing delivery in Japan". In A. E. Piroozfar, F. T. Piller (2013) *Mass Customisation and Personalisation in Architecture and Construction: A Compendium of Customer-centric Strategies for the Built Environment*, pp. 164–173. London & New York: Routledge/Taylor & Francis Group.

Nomura, T. (2012) "How people can accept robots: from the perspective of psychological experiments and social survey". Invited lecture at 2nd International Symposium on Biofied Buildings, Keio University, Tokyo, February.

Ohno, T. (1978) *Toyota Production System – Beyond Large Scale Production*. (Translated by Productivity Press, 1988). Portland: Productivity Press. (Original Japanese edition: *Toyota seisan hoshiki*, published by Diamond, Inc., Tokyo).

Osamu, S. (1994) "A history of tatami". *Chanoyu Quarterly*, 77:7–27.

Pearson, C., Delatte, N. (2005) "Ronan point apartment tower collapse and its effect on building codes". *Journal of Performance of Constructed Facilities*, 19(2):172–177.

People's Daily Online, PDO (2005) "China to push forward urbanization steadily". http://english.peopledaily.com.cn (Accessed July 2014).

Piller, F. T. (2006) *Mass Customization – ein wettbewerbsstrategisches Konzept im Informationszeitalter* (4th ed.). Wiesbaden: Deutsche Universitäts Verlag.

Pine, B. J., Gilmore, J. H. (1999) *The Experience Economy – Work Is Theatre and Every Business a Stage*. Boston: Harvard Business Review Press.

Powers, A. (2007) *Britain (Modern architectures in history)*. London: Reaktion Books.

Prefab Club. Japanese Prefabrication Construction Suppliers and Manufacturers Association (Prefab Club). http://www.purekyo.or.jp/ (Accessed 19 November 2013).

Prischow, G., Dalacker, M., Kurzm J., Zeiher, J. (1994) A mobile robot for on-site construction of masonry". In *Intelligent Robots and Systems' 94. Advanced Robotic Systems and the Real World, IROS'94. Proceedings of the IEEE/RSJ/GI International Conference* (Vol. 3, pp. 1701–1707). IEEE.

Purohit, K., Sharma, C. S. (2002) *Design of Machine Elements*. New Delhi, India: PHI Learning Pvt. Ltd.

Reynolds, M. J. (2001) *Maekawa Kuino and the Emergence of Japanese Modernist Architecture*. Berkely and Los Angeles: University of California Press.

Ritchie, T. A. (1980) "A history of the tunnel kiln and other kilns for burning bricks". *Bulletion of the Association for Presesrvation Technology*, S. 47–61.

Rizvi, G. M., Pop-Iliev, R., Parky, C. B. (2002) "A novel system design for continuous processing of plastic/wood-fiber composite foams with improved cell morphology". *Journal of Cellular Plastics*, 38(5):367–383.

Robertson, D. S. (1929) *Greek and Roman Architecture*. Cambridge: Cambridge University Press.

Russell, C. S., Vaughn, W. J. (2013) *Steel Production: Processes, Products, and Residuals*. Oxon, UK: Routledge.

Sakao, T., Lindahl, M. (2009) *Introduction to Product/Service-System Design*. London: Springer Science+Business Media.

Sakumoto, Y. (1997) *Steel-Framed Housing – Present Situation and Forecast for Its Wider Application in Japan*. Tokyo: The Kozai Club.

Sanyo Homes (2015) http://www.sanyohomes.co.jp/en/company/ (Accessed 12 February 2015).

Schaeffer, R. E. (1992) *Reinforced Concrete: Preliminary Design for Architects and Builders*. New York: McGraw-Hill.

Schleuning, J. (2000) "Die Siedlung Römerstadt in Frankfurt a. M. von Ernst May". *Diplomarbeiten Agentur*.

Schmandt-Besserat, D. (1977) "The earliest uses of clay in Syria". *Expedition: The Magazine of the University of Pennsylvania*, 19(3):28–42.

Schmieder, M., Thomas, S. (2005) *Platformstrategien und Modularisierung in der Automobilentwicklung*. Aachen: Shaker Verlag.

Scottish Development International (SDI) (2015) www.sdi.co.uk (Accessed 12 February 2015).

Sekisui Chemicals (2015) http://www.sekisuichemical.com/about/history/ (Accessed 12 February 2015).

Sekisui House (2015) http://www.sekisuihouse.co.jp/english/info/history.html (Accessed 12 February 2015).

Shingo, S. (1982) *Study of Toyota's Production System from an Engineering Viewpoint*. Tokyo: Japan Management Association (JMA).

Short, J. R. (1982). *Housing in Britain: The Post-war Experience*. London: Routledge.

Steiger, R. W. (1992) "Going to sea in concrete". *Concrete Construction*, 470–480.

Stephens, M. P., Meyers, F. E. (2013) *Manufacturing Facilities Design and Material Handling*. West Lafayette, IN: Purdue University Press.

Stordeur, D., Helmer, D., Jamous, B., Khawam, R., Molist, M., Willcox, G. (2007) "Le PPNB de Syrie du sud à travers les découvertes récentes à Tell Aswad". In *Hauran V La Syrie du Sud du Néolithique à L'Antiquité Tardive Recherche, s Récentes Actes du Colloque de Damas*, Vol. 1, pp. 41–68.

Sutherland, R. J. M., Humm, D., Chrimes, M. (eds.) (2001) *Historic Concrete: Background to Appraisal*. Thomas Telford.

Tama Home (2015) http://www.tamahome.jp/english/corporate-profile/corporate-profile (Accessed 12 February 2015).

Tamke, M., Thomsen, R. (2009) "*Digital wood craft*". In *Proceedings of the 13th International CAAD Futures Conference*. Montréal: Les Presses de l'Université de Montréal.

Taylor, S. (2011) "Offsite production in the UK construction industry – prepared by HSE: A brief overview". Report Prepared by StephenTaylor, Construction Engineering Specialist Team. London: HSE.

Thomas, J. C. (1995) *Twentieth-Century Building Materials: History and Conservation*. New York: McGraw-Hill.

Thomopoulos, N. (2014) *Assembly Line Planning*. New York: Springer Science+Business Media.

Toyota (2015) http://www.toyota-global.com/company/history_of_toyota/75years/data/business/housing/toyotahome.html (Accessed 12 February 2015).

UNECE. (2005) "European Forest Sector Outlook Study 1960-2000-2020. Main report". In *Geneva timber and forest study paper 20*. Geneva, Switzerland: United Nations Publications.

VW. Production Process in the Factory Gäserne Manufaktur http://www.glaesernemanufaktur.de/ (Accessed 19 November 2013).

Wanheim, T. (2004) "Design of forming processes, sheet metal bending". In G. E. Totten, K. Funatani, X. Xie (eds.), *Handbook of Metallurgical Process Design*, pp. 23–46. New York: Marcel Dekker.

Waters, T. F. (2002) *Fundamentals of Manufacturing for Engineers*. London: Taylor & Francis.

Weber, W., Rabaery, J., Aarts, E. (2005) *Ambient Intelligence*. Berlin: Springer Science+Business Media.

WeberHaus. Unternehmen, Häuser, Energie & Technik. www.weberhaus.de (Accessed 19 November 2013).

Weckenmann (2015) http://www.weckenmann.com/en (Accessed 12 February 2015).

Wemger, K. F. (1984) *Forestry Handbook*. New York: John Wiley & Sons.

Wentworth, I. (1982) U.S. Patent No. 4364984. Washington, DC: U.S. Patent and Trademark Office.

Wichert, R., Eberhardt, B. (2011) *Ambient Assisted Living*. Berlin: Springer Science+Business Media.

Wißnet, A. (2007) *Roboter in Japan: Ursachen und Hintergründe eines Phänomens'*. Munich: Iudicium-Verlag.

World Steel Association. http://www.worldsteel.org/ (Accessed 25 November 2013).

Yamaguchi, K., Matsufuji, Y., Koyama, T., Koyamada, H. (2005) "Strength evaluation of friction-resistant type dry-masonry structure for realizing reuse". In *The 2005 World Sustainable Building Conference Tokyo, 27–29 September 2005*.

Yoshida, M., Kubota, Y., Yokota, Y. (2005) "Mechanism of the Man-Nen Dokei: a historic perpetual cronometer; Part 1: Wadokei, a Japanese traditional clock". *Proceedings of Japan Society of Mechanical Engineers, Vol. 5*, pp. 51–52.

Zhou, Z., Ou, G., Hang, Y., Chen, G., Ou, J. (2009) "Research and development of plastic optical fiber based smart transparent concrete". In *Proceedings of SPIE 7293, Smart Sensor Phenomena, Technology, Networks, and Systems, San Diego*.

Index

Printed in the United States
by Baker & Taylor Publisher Services